BACTERIAL BIOPOLYMERS

BACTERIAL BIOPOLYMERS

Bolin Kumar Konwar, PhD

AAP | APPLE ACADEMIC PRESS

First edition published 2024

Apple Academic Press Inc.
1265 Goldenrod Circle, NE,
Palm Bay, FL 32905 USA
760 Laurentian Drive, Unit 19,
Burlington, ON L7N 0A4, CANADA

CRC Press
6000 Broken Sound Parkway NW,
Suite 300, Boca Raton, FL 33487-2742 USA
4 Park Square, Milton Park,
Abingdon, Oxon, OX14 4RN UK

Apple Academic Press exclusively co-publishes with CRC Press, an imprint of Taylor & Francis Group, LLC

Library and Archives Canada Cataloguing in Publication

Title: Bacterial biopolymers / Bolin Kumar Konwar, PhD.
Names: Konwar, B. K., author.
Description: First edition. | Includes bibliographical references and index.
Identifiers: Canadiana (print) 20230142176 | Canadiana (ebook) 20230142230 | ISBN 9781774912058 (hardcover) | ISBN 9781774912065 (softcover) | ISBN 9781003331636 (ebook)
Subjects: LCSH: Poly-beta-hydroxyalkanoates. | LCSH: Biopolymers. | LCSH: Bacteria.
Classification: LCC QR92.P58 K66 2023 | DDC 572—dc23

Library of Congress Cataloging-in-Publication Data

Names: Konwar, B. K., author.

Title: Bacterial biopolymers / Bolin Kumar Konwar.

Description: Palm Bay, FL : Apple Academic Press, 2023. | Includes bibliographical references and index. | Summary: "Bacterial biopolymers are a competitive new area of research, having potential application in agriculture, petroleum exploration, and use as biodegradable plastics. This new volume provides comprehensive coverage of bacterial biopolymers. The book elucidates on the main classes of bacterial biopolymers-polysaccharides and polyesters or polyhydroxyalkanoates (PHAs)-along with their definition, classification, detection, extraction, characterization, fermentation process, structure and properties, their applications and uses, functions, and more"-- Provided by publisher.

Identifiers: LCCN 2023001213 (print) | LCCN 2023001214 (ebook) | ISBN 9781774912058 (hardcover) | ISBN 9781774912065 (paperback) | ISBN 9781003331636 (ebook)

Subjects: LCSH: Biopolymers. | Poly-beta-hydroxyalkanoates. | Bacteria.

Classification: LCC TP248.65.P62 K65 2023 (print) | LCC TP248.65.P62 (ebook) | DDC 668.9--dc23/eng/20230302

LC record available at https://lccn.loc.gov/2023001213

LC ebook record available at https://lccn.loc.gov/2023001214

ISBN: 978-1-77491-205-8 (hbk)
ISBN: 978-1-77491-206-5 (pbk)
ISBN: 978-1-00333-163-6 (ebk)

About the Author

Bolin Kumar Konwar, PhD, was a Dean in the School of Science and Technology and a Professor in the Department of Molecular Biology and Biotechnology, Tezpur University (Central) India. He also headed the ONGC-sponsored Centre for Petroleum Biotechnology, DBT (Govt. of India)-sponsored MSc biotechnology program, and the Bioinformatics Infrastructure Facility (BIF). He was formerly the Vice Chancellor at Nagaland University (Central), Nagaland, India, and also served at Assam Agricultural University, Jorhat, from 1984-1995, as Lecturer, Assistant Professor, and Associate in the Department of Plant Breeding and Genetics. He was Senior Scientist and Head of Botany and Biotechnology at the Tocklai Experimental Station, Tea Research Association, Jorhat, Assam, India, from 1995–2002. He carried out research projects of national importance as the principal investigator and supervised many students on research projects, with 15 students earning PhDs to date. He and his research group have so far deposited 11 gene (DNA) sequences in gene banks, and his research group has published over 200 research papers in international journals, seminars, and conference proceedings, and have presented over 90 research papers at seminars and conferences in India and abroad. Dr. Konwar has published more than 300 articles on popular science, environment, biotechnology, history, national integration, higher education, research needs, etc., in various Assamese magazines and newspapers, as well as more than 40 scientific articles in national English magazines in addition to eight books, three booklets, and 25 book chapters. The books written by by Dr. Konwar are *Lipase: An Industrial Enzyme through Metagenomics and Bacterial Biosurfactants: Isolation, Purification, Characterization, and Industrial Applications (Apple Academic Press); Prospects of Microbe and Medicinal Plant Resources; Medicinal Plant Repertoire: A Perspective of Biogeographical Gateway of India; Joimati–Gadapani: Ek Amar Dampatya Premgatha; Deshapremi Pariyal: Barbaruah Barphukan (in Assamese), and The Struggle with a Hill University: Administering Higher Education in a Remote Hilly Terrain.*

Dr. Konwar earned a DIC in Microbiology and PhD in Plant Biotechnology at the Imperial College of Science, Technology and Medicine, University of London.

Contents

Abbreviations

AFM	atomic force microscopy
AuNPs	gold nanoparticles
CDW	cellular dry weight
CLSFM	confocal laser scanning fluorescence microscopy
CTAB	cetyltrimethylammonium 46 bromide
CV	cyclic voltametric
DB	Dry biomass
DSC	differential scanning calorimetry
ECS	extruded corn starch
ERB	extruded rice bran
FAME	fatty acid methyl esters
FTIR	Fourier-transform-infrared
GC	gas chromatography
GPC	gel permeation chromatography
H2S	hydrogen sulfide
HA	hydroxyapatite
HBE	hyper-branched epoxy
KBr	potassium bromide
KCDL	kitchen chimney dump lard
MHBE	modified hyperbranched epoxy
MR	methyl red
MR-VP	methyl red-Voges-Proskauer
MSM	minimum salt medium
MW	molecular weight
NB	nutrient broth
NPs	nanoparticles
PCL	polycaprolactone
PGA	polyglycolic acid
PHA	polyhydroxyalkanoate
PHB	polyhydroxybutyrate
PHBHHx	poly(3-hydroxybutyrate-co-3-hydroxyhexanoate)
PHB-HV	poly-3-hydroxybutyrate-co-3-hydroxyvalerate

PL	photoluminescence
PLA	polylactic acid
PVA	polyvinyl alcohol
PVP	polyvinylpyrrolidone
RT-PCR	reverse transcription and polymerase chain reaction
SD	sabouraud dextrose
SDS	sodium dodecyl sulfate
SEM	scanning electron micrograph
SIM	sulfur indole motility
SNP	silver nanoparticles
ST	Surface tension
TEM	transmission electron microscope
TGA	thermogravimetric
THF	tetrahydrofuran
TIC	total ion chromatogram
VFAs	volatile fatty acids
XRD	X-ray diffraction

Foreword

The world is becoming increasingly concerned over the harmful effect of petrochemical-derived plastic materials. This has necessitated the development of biodegradable plastic, retaining the physical and chemical properties of conventional synthetic plastics. The existing methods to degrade synthetic plastics are inefficient; incineration is an alternative option, but this is not only costly and causes the emission of harmful chemicals like hydrogen chloride and hydrogen cyanide during incineration. These problems have led to the idea of greener, eco-friendly, biodegradable plastics. Biopolymers are a diverse and versatile class of materials that have potential applications in virtually all sectors of the economy. They are naturally biodegradable polymers that are synthesized and catabolized by various microorganisms, plants, and animals, and they do not cause toxicity in the host and have certain advantages over petroleum-derived plastics. Biopolymers are nucleic acids, proteins, polysaccharides, polyphosphates, polyphenols, and poly-sulfates. They are prone to enzymatic degradation as they are the product of enzymatic polymerization for their synthesis.

Beijerinck first observed lucent granules of PHA in bacterial cells in 1888, and the French scientist Maurice Lemoigne 1926 first discovered PHA in *Bacillus megaterium*. Several bacteria synthesize and accumulate PHA as a sink for carbon and energy in limiting nutrients in the presence of excess carbon. More than 90 different genera of archaea and eubacteria are known to accumulate PHAs. A wide variety of PHA copolymers are isolated from bacteria, like many cyanobacteria found in different environmental samples, including marine sediments. PHAs are the most versatile group of naturally occurring biodegradable and biocompatible polymers that belong to the aliphatic polyester family. In the cytoplasm of various microorganisms, including bacteria, in the presence of an excess of carbon source and nutrient limiting conditions, PHAs accumulate as water-insoluble inclusion bodies for storing the compounds of carbon and energy during the imbalanced growth. Certain bacteria, however, produce them without being subjected to any kind of nutritional constraints, such as *Alcaligenes latus*. PHAs can contribute up to 90% (w/w) of the cell dry weight. The number per cell and size of inclusion bodies of PHA varies among different species. Members of the PHA family can exist as homopolymers of alkanoic hydroxy acids, as well as copolymers of two or more hydroxy alkanoic acids.

For commercializing PHA, various efforts have been devoted to reducing the cost of production with the development of superior bacterial strains using recombinant technology, more efficient fermentation, and recovery processes. A large number of bacteria like *Axobacter, Azospirillum, Beijerinckia, Clostridium, Halobacterium, Methylocystis, Rhizobium, Rhodospirillum, Cyanobacteria, Paracoccus denitrificans, Pseudomonas stutzeri, P. aeruginosa*, recombinant *Bacillus subtilis, Delftia acidovorans, Azotobacter chroococcum, A. vinelandii, Streptomyces aureofaciens, Aeromonas hydrophila*, etc., are known to produce biopolymer.

PHA synthesis occurs intracellularly in multiple cytoplasmic inclusion bodies that are generally 200–500 nm in size and surrounded by a membrane to which proteins are bound. PHA granules may be surrounded by a phospholipid monolayer that contains specific granule-associated proteins such as (i) PHA synthase (PhaC), (ii) PHA depolymerases and 3HB-oligomer hydroxylase (PhaZi), (iii) phasins (PhaPs), which are thought to be the major structural proteins of the membrane surrounding the PHA-containing inclusion bodies, and (iv) the regulator of phasin expression (PhaR) and also cytosolic proteins that may be non-specifically attached to the granules via hydrophobic interactions. Based on the number of carbon atoms in the monomeric unit, PHAs are classified into (i) short chain length PHAs (scl-PHAs; 3–5 C-atoms) and (ii) Medium chain length PHAs (Mcl-PHA; 6 or more C-atoms).

Preface

The introduction of synthetic polymer materials has turned out to be a vital part of our contemporary life due to their various attractive properties, which have molded society in many ways that make life both easier and safer. In the present world, the most extensively used synthetic polymers are petroleum-derived synthetic plastics. Plastic materials are composed of different synthetic or semi-synthetic products and can be manipulated chemically to have better strength and shapes as per our interests. Chains of carbon atoms alone or in combination with oxygen, sulfur, or nitrogen make up the majority of the polymers. Many repeating units together, being part of the chain on the main path, comprise the backbone of plastic polymers. They have a high molecular weight (MW) ranging from 50,000–1,000,000 Da., and have many desirable properties such as high chemical resistance, variable elasticity, malleability, non-toxicity, and hence are popular in many durables, disposable goods, and packaging materials of our day-to-day life. Synthetic materials leave harmful imprints on the environment. The main cause of the inconvenience of synthetic plastics towards the environment is their disposal as they are recalcitrant to microbial degradation. Because of the durability of synthetic plastics and their disposable usage, wastes contribute to an array of environmental problems.

Polylactic acid (PLA) is biodegradable thermoplastic aliphatic polyester derived from the fermentation of carbohydrate feedstocks such as maize, wheat, or agricultural wastes like whey, molasses, green juice, etc. Bacterial fermentation of sugars is used to produce lactic acid monomers, oligomerized and catalytically dimerized, to form lactic acid polymers. In recent years, considerable attention has been paid to biopolymers. Polyhydroxyalkanoates (PHAs) belong to the family of microbial polyesters that accumulate as storage material in microbial cells under stress conditions. They are aliphatic polyesters of various hydroxyalkanoates synthesized by microorganisms as storage of carbon and energy. Inclusion bodies in the microbial cytoplasm act as the storehouse of PHA in the excess carbon source when an essential nutrient like nitrogen or phosphorus is limited.

PHAs extracted from the bacterial cells are strong biodegradable materials having properties like various synthetic thermoplastics and elastomers, from polypropylene to synthetic rubber.

The use of PHAs in a wide range of applications has been hampered mainly due to their high cost of production as compared to petrochemical-based polymers. The production from renewable carbon sources, such as agro-horticultural wastes of corn, cassava, etc., could be economically important in view of environmental gains that would result from their application.

The major problem behind PHAs' commercial production and application is the high cost. Bacterial PHA production is 5–10 times more expensive than petroleum-derived polymers such as polyethylene and polypropylene. The ability of the bacteria to utilize cheap carbon sources is important so as to reduce the cost. Therefore, PHA production is mainly focused on renewable resources where bacteria can utilize waste generated from food, agriculture, and industrial processes and fatty acids as carbon sources.

The bacterial cell biomass is harvested from the culture broth using centrifugation, filtration, or flocculation-centrifugation for the recovery of PHA. The lyophilized harvested cells are extracted with the methods like solvent extraction or non-PHA biomass digestion. Enzymes such as alcalase, neutrase, lecitase, and lysozyme are used for the digestion, followed by washing with sodium dodecyl sulfate (SDS), an anionic surfactant to solubilize cellular material non-P(3HB). The solubilized and non-solubilized cell compounds are centrifuged following washing with anionic surfactant and flocculation. The physical properties of biopolymers are mainly dependent on the chemical formulation and structure of the polymer itself. PHB is an optically active, compact helical structure, which is most like that of polypropylene, which also has a compact configuration. Characteristics of biopolymers are almost like those of conventional plastics like polypropylene. The breakdown of complex PHA compounds into simpler compounds due to the action of microorganisms without the release of hazards to the environment is desirable; different biodegradations occur in the environment, such as hydro-biodegradation, photo-biodegradation, and oxo-biodegradation.

CHAPTER 1

Introduction

The introduction of synthetic polymer materials has turned out to be a vital part of our contemporary life due to their various attractive properties, which have molded society in many ways that make life both easier and safer. Most synthetic polymers are petroleum-derived plastics that are composed of different synthetic or semi-synthetic products and can be manipulated chemically to have better strength and shapes as per interests. The vast majority of the polymers are based on chains of carbon atoms alone or with oxygen, sulfur, or nitrogen as well. The backbone of plastic polymers is that part of the chain on the main "path" linking many repeating units together. They have a high molecular weight (MW) ranging from 50,000 to 1,000,000 Da (Parvizi, 2010). The molecular structure of plastics has allowed them to become an indispensable part of the present century having many desirable properties such as high chemical resistance, variable elasticity, malleability, and non-toxicity, and hence popular in many durables, disposable goods, and packaging materials. But the synthetic polymers are harmful to the environment. The main cause of the inconvenience of plastics to the environment is their disposal, as they are recalcitrant to microbial degradation. The chemical building blocks that make plastics so versatile are the components that might harm people and the environment. The durability of plastics, as well as their disposable usage, contribute to serious environmental problems:

i. Chemicals used in plastics are absorbed by human bodies, and some of these compounds affect normal hormonal behavior affecting health. Also, these chemicals are known to leach from

Bacterial Biopolymers. B. K. Konwar, PhD (Author)

landfill into soil and waterbodies, causing aquatic contamination and often ingested by marine animals, injuring, and poisoning aquatic life.

ii. Plastic buried deep in landfills can leach harmful chemicals that spread into groundwater, and those remaining on land and floating on water bodies impair ugly looks all around.

iii. A sizeable volume of petrol crude is used as a feedstock to make plastics the world over, and a similar amount is consumed as energy in the process.

In recent years, there has been public outcry over the harmful effect of petrochemical-derived synthetic plasticware on the environment. This has prompted many countries to start serious research in developing biodegradable plastics retaining the same physical and chemical properties as synthetic plastics. According to an estimate, around 8% of the world's oil production is used to make plastics, and the production of plastic has crossed 8.30 billion tons per annum (Siracusa et al., 2020). This problem is compounded by the fact that the resource for crude oil is also decreasing.

Existing methods to degrade synthetic plastics are inefficient. An incineration is an alternate option in dealing with non-degradable plastics, but in addition to its high cost, harmful chemicals like hydrogen chloride and hydrogen cyanide are emitted during incineration (Atlas, 1993). Therefore, the idea of substituting non-degradable synthetic plastics with greener, eco-friendly biodegradable plastics has drawn interest from both the academic and commercial world. This alternative has come in the form of biodegradable plastic materials.

KEYWORDS

- **biodegradable plastics**
- **incineration**
- **microbial degradation**
- **petrochemical-derived synthetic**
- **plasticware**
- **polymer materials**

REFERENCES

Atlas, R. M., (1993). *Microbial Ecology: Fundamentals and Applications* (3rd edn.). Benjamin-Cummings Publishing Company Inc. Redwood City, CA.

Parvizi, J., (2010). Polyethylene. In: High Yield Orthopedics (pp. 1–15).

Siracusa, V., et al., (2020). Bio-polyethylene (Bio-PE), bio-polypropylene (bio-PP), and bio-poly(ethylene terephthalate) (Bio-PET): Recent developments in bio-based polymers analogous to petroleum-derived ones for packaging and engineering applications. *Polymers*, *12*(8), 1641.

CHAPTER 2

Biopolymers and Their Classification

Polymers of biological origin are not only synthesized and catabolized by microorganisms but also by animals and plants. Microorganisms, with their enzymatic reactions, can easily degrade biopolymers giving the least chance for the accumulation of toxic chemicals in the environment, thereby allowing restoration of the ecosystem. Biopolymeric products formed through the aggregation of monomeric units cause various functions in the producing organisms. Biochemical compounds like polyphosphates, polyphenols, polysaccharides, polysulfates, polyesters, nucleic acids-DNA and RNA, and proteins represent biopolymers. They can be extensively used in various industrial and other economic sectors. They are superior to synthetic polymers in respect of human health and eco-friendliness. On the basis of their synthesis, they can be broadly classified into the following four groups:

- **Type A:** Those isolated from agro-biomass or agro-resources such as starch, cellulose, etc.
- **Type B:** Those obtained from microbes as well as from genetically modified bacteria PHA and cellulosic mass of bacteria.
- **Type C:** Polymers conventionally and chemically synthesized from renewable biobased monomers obtained from agro-resources, the poly lactic acid: biodegradable polyesters polymerized from lactic acid monomers.
- **Type D:** Those monomers and polymers obtained conventionally by chemical synthesis specifically from petroleum. Biodegradable polymers like aliphatic-aromatic co-polyesters. Though

Bacterial Biopolymers. B. K. Konwar, PhD (Author)

such polymers are made from synthetic components, but can be composted and biodegraded.

2.1 POLYMERS FROM AGRO-BIOMASS

The biopolymers of Type 1 are the most common in the environment and are extracted from marine and agricultural animals and plants, such as:

1. **Starch:** It is the storage polysaccharides of cereals, tubers, and legume plants and widely available agro-resources suitable for a variety of industrial uses. Glucose molecules joined by glycosidic bonds mainly lead to the formation of starch which is competitive with petroleum and used in several processes of preparing compostable plastics.
2. **Cellulose:** Biological matter is most abundantly comprised of cellulosic constituents which is the principal component of plant cell walls. They are almost linear polymer of anhydroglucose and their regular structure and tendency to form hydrogen-bonded crystalline microfibrils and fibers make them good candidates as packaging material.
3. **Chitosan:** The exoskeleton crustaceans and fungi cell walls are mostly made of chitosan, a natural biodegradable polysaccharide. Chemically it is the linear polysaccharide is chemically composed of randomly available-(1-4)-linked D-glucosamine (deacetylated) and N-acetyl-D-glucosamine (acetylated) units. It has several uses in commercial and biomedical fields. The antimicrobial property makes it to be good packaging material; it is also used for seed treatment, plant growth enhancer, and eco-friendly biopesticide.

2.2 PROTEINS: CASEIN, KERATIN, COLLAGEN, GLUTEN, SOY

Casein, keratin, and collagen are natural animal-derived proteins. Gluten and soy protein are present in plants like wheat, corn, and legumes. They are insoluble in water and difficult to process.

2.3 LIGNIN

Lignin is hydrophobic aromatic complex compound most commonly found in wood and integral part of plants' secondary cell wall (Lebo et al., 2000). It is an abundant organic polymer on earth next to cellulose. It is unusual due to heterogeneity, defined primary structure, and covalently linked to hemicellulose as well as could crosslink with the different plant polysaccharides resulting mechanical strength to the cell wall and to the plant as a whole. It plays an important role in water transport in plant stem and decomposes slowly through decomposition of dead vegetation, contributing mostly to humus.

2.4 HYALURONIC ACID

It is specifically a mucopolysaccharide carbohydrate found in all vertebrate tissues and also extracellularly in a variety of bacterial polysaccharides. It plays an important physiological role in reproductive cells by providing a protective matrix, in the lymphatic system acts as a regulator, and a joint lubricating fluid. This characteristic makes it most beneficial substance in orthopedics and eye surgery.

2.5 PRODUCTION OF PHA IN HIGHER ORGANISMS

2.5.1 YEASTS

The yeast (*Saccharomyces cerevisiae*) cells are spherical or ellipsoidal with a diameter of 25 m. The thick (100,200 nm) envelope-like yeast cells with poor permeability to water-soluble molecules possess an average mol weight > 700 and radius >1 nm. In yeast cell walls, it is major structural constituent polysaccharides (81%), mainly 1,3-glucans. Poirier (1992) reported the transformation of PHA synthase from *Pseudomonas aeruginosa* bacteria with modification at the carboxy-end for peroxisomal targeting in methylotrophic yeast *Pichia pastoris*. The promoter of *P. pastoris* acyl-CoA oxidase gene expresses the PHA synthase and accumulates medium-chain-length PHA to 1% per g dry within the peroxisomes. Bohmert et al. (2000) reported production of PHAs in a transgenic yeast

Saccharomyces pombe harboring the PHB synthesis genes encoding b-ketothiolase (phbARe), acetoacetyl-CoA reductase (phbBRe) and PHB synthase (phbCRe) of *R. eutropha*. The transgenic yeast accumulated about 9% (w/w) of PHB under optimized conditions. In metabolically engineered Kloeckera yeast cells PHA accumulation can be up to 7% (w/w).

2.5.2 INSECTS

The amino polysaccharide polymer chitin abundantly occurs in nature, as building material that gives strength to the exoskeleton of insects, crustaceans, and fungal cell walls. It is converted to chitosan through enzymatic or chemical deacetylation. Shrimp and crabs mostly provide chitin being an abundant byproduct of food-processing industry and the biopolymer is used in biomedical applications. Its synthesis and degradation need enzymatic control to maintain homeostasis (Elieh-Ali-Komi and Hamblin, 2016). It has been reported by Williams et al. (1992) reported a novel pathway for the synthesis of P(3HB) by engineering two genes into insect *Spodoptera frugiperda* cells using baculovirus vector. The system expresses a dehydrase-domain mutant rat fatty acid synthase cDNA and the phbC gene encoding PHA synthase from *A. eutrophus*. Approximately 1 mg of PHB was isolated from a one-liter culture of these cells corresponding to 0.16% of cell dry weight.

2.5.3 PLANTS

Plants produce a range of biopolymers for purposes such as maintenance of structural integrity, carbon storage, and defense against pathogens and desiccation. Several of these natural polymers are used by humans as food and materials, and increasingly as an energy carrier. Plants are able to produce several useful chemicals at a low cost as compared to bacteria or yeast. The commercialization of plant-PHA needs development of transgenic plants with the addition of high product yields. The plant-based production of PHAs, silk, elastin, collagen, and cyanophycin with the synthesis of poly [(R)-3-hydroxybutyrate] (PHB) with potential commercial applications in plastics, and food markets. The improved production of PHB could be achieved the expression of transgenes, reduction of endogenous enzymes activity and cloning of genes to increase carbon flow to polymer (Snell et al., 2015). Transgenic plants could produce PHAs from photosynthetically fixed CO_2 and water which after disposal degrade back to CO_2 and water. Therefore, it might be possible to produce PHA at a similar low cost which is comparable to those of other biopolymers already obtained from plants. Arabidopsis thaliana was the first targeted model for transgenic studies in plants. Poirier (1992) reported synthesis of PHA in plants was initially explored by the expression of PHA biosynthetic genes of the bacterium *R. eutropha* in the well-studied plant *A. thaliana.* But less (0.1%) PHB accumulation was achieved from the plant. Recently all three genes necessary for PHB biosynthesis were transformed to *A. thaliana* in a single transformation event. These plants accumulated more than 4% of their fresh weight (approximately 40% of their dry weight) of PHB in leaf chloroplasts. Bohmert et al. (2000) reported the novel perspective on the use of PHA synthesis in cotton fiber cells by the expression of the *R. eutropha* PHB biosynthetic pathway. Analysis of the transgenic fibers showed PHB accumulation up to 0.3% of dry weight exhibiting better insulating properties. Moreover, PHA biosynthetic genes also have been expressed in some agricultural crops such as *Brassica napus*, *Gossypium hirsutum*, Nicotiana tabacum, Solanum tuberosum, and *Zea mays* (Slater et al., 1999). By transforming the threonine deaminase gene from *E. coli* and PHB biosynthetic genes from *R. eutropha*, PHBV were produced by engineered Arabidopsis and Brassica plants in their leaves and seeds, respectively. A gene construct of plastid transformation in *Nicotiana tabacumto* to produce PHB was reported by Bohmert-Tatarev et al. (2011).

KEYWORDS

- **biopolymers**
- **chitosan**
- **exoskeleton**
- **hydrophobic aromatic**
- **poly((R)-3-hydroxybutyrate)**
- ***Saccharomyces cerevisiae***

REFERENCES

Akar, A., et al., (2006). Accumulation of polyhydroxyalkanoates by *Microlunatus phosphovorus* under various growth conditions. *J. Ind. Microbiol. Biotech., 33*, 215–220.

Angelova, N., & Hunkeler, D., (1999). Rationalizing the design of polymeric biomaterials. *Trends Biotechnol., 17*, 409–421.

Bohmert, K., et al., (2000). Transgenic *Arabidopsis* plants can accumulate polyhydroxybutyrate to up to 4% of their fresh weight. *Planta, 211*, 841–845.

Bohmert-Tatarev, K., et al., (2011). High levels of bioplastic are produced in fertile transplastomic tobacco plants engineered with a synthetic operon for the production of polyhydroxybutyrate. *Plant Physiol., 155*, 1690–1708.

Elieh-Ali-Komi, D., & Hamblin, M. R., (2016). Chitin and chitosan: Production and application of versatile biomedical nanomaterials. *Int. J. Adv. Res. (Indore), 4*(3), 411–427.

Hahn, J. J., et al., (1999). Peroxisomes as sites for synthesis of polyhydroxyalkanoates in transgenic plants. *Biotechnol. Prog., 15*, 1053–1057.

Houmiel, K. L., et al., (1999). Poly (b-hydroxybutyrate) production in oilseed leucoplasts of *Brassica napus*. *Planta, 209*, 547–550.

John, M. E., & Keller, G., (1996). Metabolic pathway engineering in cotton: Biosynthesis of polyhydroxybutyrate in fiber cells. *Proc. Natl. Acad. Sci. USA, 93*, 12768–12773.

Kim, D. Y., & Rhee, Y. H., (2003). Biodegradation of microbial synthetic polyesters by fungi. *Appl. Microbiol. Biotechnol., 61*, 300–308.

Lebo, Jr. S. E., Gargulak, J. D., & McNally, T. J., (2001). *Lignin Kirk–Othmer Encyclopedia of Chemical Technology.* John Wiley and Sons Inc.

Poirier ,Y., et al., (1992). Polyhydroxybutyrate, a biodegradable thermoplastic, produced in transgenic plants. *Science, 256*, 520–523.

Slater, S., et al., (1999). Metabolic engineering of *Arabidopsis* and brassica for poly (3-hydroxybutyrate-co-3-hydroxyvalerate) copolymer production. *Nat. Biotechnol., 17*, 1011–1016.

Snell, K. D., et al., (2015). Production of novel biopolymers in plants: Recent technological advances and future prospects. *Current Opinion in Biotech, 32*, 68–75.

Tokiwa, Y., & Calabia, B. P., (2004). Degradation of microbial polyesters. *Biotechnol Lett., 26*, 1181–1189.

William, M. D., et al., (1992). Production of a polyhydroxyalkanoate biopolymer in insect cells with a modified eucaryotic fatty acid synthase. *Appl. Envt. Microb., 62*(7), 2540–2546.

CHAPTER 3

Bacterial Polymers

3.1 XANTHAN

The gum is a complex microbial polysaccharide derived from the bacterial coat of *Xanthomonas campestris*; glucose, sucrose, or lactose fermentation by the bacteria could cause production. The remarkable properties of xanthan gum make it an attractive polymer for different industrial and biological purposes.

3.2 PULLULAN

It is a linear polysaccharide extracellularly produced by several species of yeast, specifically by *Aureobasidium pullulans*. Its long chain consists of maltotriose units connected with each other by a glycosidic bond. Pullulan has many uses as an industrial plastic, in the food industry.

3.3 BACTERIAL CELLULOSE

It is a straight polymer chain consisting of β-linkage bound glucose units. In addition to bacteria like *Acetobacter*, *Agrobacterium*, and *Sarcina* genera algae also contain cellulose. Its chemical and physical nature also correlates with the cellulose found in plants (Angelova and Hunkeler, 1999). Some bacteria synthesize exocellulose to form protective cellular envelopes in natural habitats. Nowadays, bacterial cellulose is widely used for a variety of commercial applications with modifications in their synthesis.

Bacterial Biopolymers. B. K. Konwar, PhD (Author)

3.4 POLYLACTIC ACID (PLA) WITH RENEWABLE BIO-BASED MONOMERS OF AGRO-RESOURCES

Polylactic acid (PLA) is biodegradable thermoplastic aliphatic polyester derived from the fermentation of carbohydrate feedstock such as maize, wheat or agricultural wastes, whey, molasses, green juice, etc. Bacterial fermentation of sugars is used to produce lactic acid monomer which is ultimately oligomerized and catalytically dimerized to form lactic acid polymer (Zinn et al., 2001; Williams and Martin, 2002). PLA is currently used in several biomedical applications, agricultural, and manufacturing uses as well.

3.5 CHEMICALLY SYNTHESIZED PETROLEUM-BASED MONO- AND POLYMERS

3.5.1 POLYGLYCOLIC ACID (PGA)

Polyglycolic acid (PGA) is a simplest linear, aliphatic, thermoplastic, biodegradable polymer, produced through polycondensation or ring-opening polymerization of glycolic acid monomer molecule. Though it is considered as a tough fiber-forming polymer, the hydrolytic instability due to the presence of ester linkage in the backbone, it has limited use. Nowadays, polyglycolide and its copolymers polylactic-co-glycolic acid with lactic acid; polyglycolide-co-caprolactone with caprolactone; and polyglycolide-co-trimethylene carbonate with trimethylene carbonate are extensively used in synthesizing absorbable suture for biomedical application. Their degradation processes are erosive and easily hydrolyzed in an aqueous environment. After hydrolysis PGA is converted into some nontoxic product which is again metabolized into water and carbon dioxide and excreted by urine.

3.5.2 POLYCAPROLACTONE (PCL)

It is biodegradable thermoplastic polyester produced through chemical synthesis of low melting point (~ 60°C) crude oil. Caprolactone with ring-opening polymerization using catalyst like stannous octoate PCL could be prepared (Reddy et al., 2003). Since this polymer is compatible in nature, therefore it can be mixed with other material to improve its processing characteristic, quality, and biodegradability. In suitable physiological conditions, the hydrolysis of ester linkages could easily degrade PCL. Due to its lower melting temperature, high biodegradability and good compatibility, it is widely used for biomedical purposes like drug delivery, tissue grafting, etc.

3.5.3 POLYVINYL ALCOHOL (PVA)

Polyvinyl alcohol (PVA) is a water-soluble synthetic polymer prepared by the continuous hydrolysis of polyvinyl acetate in ethanol with potassium hydroxide. PVA is an odorless and tasteless, translucent, white to cream colored granular powder. Depending on its polymerization and hydrolysis degrees, it can be used for various functional purposes. It possesses high tensile strength and flexibility, emulsifying, as well as high oxygen and aroma barrier properties. It is fully degradable with quick-dissolving. Owing to pyrolysis at high temperatures PVA decomposes rapidly above 200°C.

KEYWORDS

- **bacterial cellulose**
- **polycaprolactone**
- **polyglycolic acid**
- **polylactic acid**
- **polyvinyl alcohol**
- **pyrolysis**

REFERENCES

Angelova, & Hunkeler, (1999). Rationalizing the design of polymeric biomaterials. *Trends Biotechnol., 17*, 409–421.

Reddy, C. S., et al., (2003). Polyhydroxyalkanoates: An overview. *Bioresour Technol., 87*, 137–146.

Williams, S. F., & Martin, D. P., (2002). Application of PHAs in medicine and pharmacy. In: Doi, Y., & Steinbuchel, A., (eds.), *Biopolymers, Polyesters III: Applications and Commercial Products.* Weinhein, Germany: Wiley-VCH.

Zinn, M., et al., (2001). Occurrence, synthesis and medical application of bacterial polyhydroxyalkanoates. *Adv. Drug Rev., 53*, 5–21.

CHAPTER 4

Polyhydroxyalkanoates (PHAs)

In the recent years, considerable attention has been paid to microbial biopolymers. Polyhydroxyalkanoates (PHAs) belong to the family of microbial polyesters that accumulate as storage material in microbial cells under stress conditions (Chen and Wu, 2005a, b). They are aliphatic polyesters of various hydroxyalkanoates synthesized by microorganisms as storage carbon and energy compounds and stored in inclusion bodies within the cytoplasm of the microbial cells (Steinbuchel, 2005; Lebo, Gargulak, and McNally, 2001). In the presence of excess carbon source if an essential nutrient element like nitrogen or phosphorus is limited the cultured microorganism produces PHA (Barnard and Sanders, 1989; Sudesh et al., 2000; Chen et al., 2001). PHAs extracted from the bacterial cells are the strong biodegradable materials having properties like various synthetic thermoplastics and elastomers, from polypropylene to synthetic rubber (Kadouri et al., 2005). In 1888 Beijerinck first reported the presence of PHA in bacterial cells (Berlanga et al., 2006) and Lemoigne its composition in the bacterium *Bacillus megaterium* as a homopolyester of 3-hydroxybutyric acid, known as polyhydroxybutyrate (PHB) (Williamson and Wilkinson, 1958) and it a prototype of PHA family. More than 100 different PHA monomers with attributes and different structures have been reported (Steinbuchel et al., 1995). Polymers of the PHA family are constantly increasing in number due to the continuous discovery of new homo- and copolymers which ultimately results in the availability of PHAs with a wide range of chemical structures and an assortment of properties (Anderson and Dawes, 1990; Steinbuchel, 1991). Many microorganisms could degrade these biopolymers by intracellular depolymerases (Poirier et al., 1995) and upon disposal, are converted to CO_2 and water in aerobic condition and to methane under the anaerobic condition. PHAs are the

Bacterial Biopolymers. B. K. Konwar, PhD (Author)

most versatile group of naturally occurring biodegradable and biocompatible polymers that belong to the aliphatic polyester family as shown in Figure 4.1 (Philip et al., 2007). In the cytoplasm of various microorganisms including bacteria in the presence of excess of carbon source and nutrient (like nitrogen, phosphorous, etc.), limiting conditions, PHAs accumulate as water-insoluble inclusion bodies for storing these compounds of carbon and energy during imbalanced growth (Choi et al., 2004). Bacteria like *Alcaligenes latus* produce without having any kind of nutritional stress (Barnard and Sanders, 1989). The resulting granules are coated with a layer of phospholipids and proteins. The granule-associated proteins play a major role in the synthesis of PHAs.

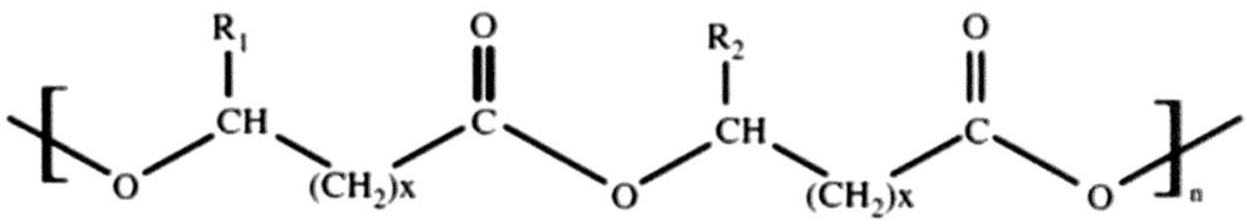

FIGURE 4.1 General structure of polyhydroxyalkanoates, R_1/R_2 = alkyl groups C_1–C_{13}, X = 4, n = 100–30,000).

Source: Reprinted with permission from Phillip et al., 2007. © John Wiley and Sons.

More than 100 different types of monomer units with different structures have been identified as constituents of the PHA (Figure 4.1). PHAs can contribute up to 90% (w/w) of the cell dry biomass. The size and number of inclusion bodies of PHA varies among different species. The monomer composition of PHA can be manipulated by changing the environment and carbon source resulting in variable structures (Lemoigne, 1926). This creates a possibility for producing different types of biodegradable polymers with an extensive range of properties. Members of the PHA family can exist as homopolymers and copolymers of two or more hydroxyalkanoic acids. These polymers are formed due to bonding of the carboxyl group of one monomer with the hydroxyl group of the next monomer through an ester linkage (Chowdhury, 1963). PHAs are synthesized by numerous Gram-positive bacteria; aerobic (cyanobacteria) and anaerobic (non-sulfur) photosynthetic bacteria as well as some archaebacteria by fermenting carbon compounds (Steinbuchel, 2005). The molecular mass of PHA is in the range of 5×104,106 Da and varies with the PHA producer. Owing to the stereospecificity of biosynthetic enzymes present inside the bacterial cell, the monomer units of the PHA polymers used to be in D-configuration (Steinbuchel, 2001; Williams, 1999). Due to low solubility, high molecular weight (MW), and specifically water insoluble property of PHAs make them an ideal carbon-energy storage material which exerts negligible osmotic pressure on the bacterial cell (Shishatskaya, 2004). Members of PHA possess polymer characteristics like those of conventional plastics such as polypropylene (Chowdhury, 1963). Other biodegradable polymers like polyglycolic, polylactic acids (PLAs) and starch-polyethylene tend to have prospects. As PHAs are biodegradable and immunologically inert, they have promising future applications, particularly in medical-related fields, despite their expensive production. The use of PHA in a wide range of applications has been hampered mainly due to their high cost of production as compared to petrochemical-based polymers (Mergaert, 1993; Brown, 1996). The use of agro-horticultural wastes in the production of PHA could be economic in view of environmental from PHA application. For commercializing PHA, efforts have been made to reduce the cost of production by using superior bacterial strains most likely recombinant microbial strains and more efficient fermentation and recovery processes (Tsuji, 2002; Flieger et al., 2003, Lee, 1995). Analysis of the entire production and recovery process of PHA would lead to design efficient methods of PHA production, and to evaluate the approximate price of PHA produced on a commercial scale. But there is need to enhance the physical properties of PHAs (Madison and Huisman, 1999). Industrial producers are currently working towards decreasing the cost of these biopolymers by increasing the volumetric production capacity of fermenter systems and improving process technology, especially the downstream processing. The open mixed microbial culture facilitates the use of mixed substrates since the microbes can effectively adapt to changes in substrate, this also reduces sterilization and sterile fermentation systems.

4.1 OCCURRENCE OF PHA

The first discovery of PHA was monitored by a French scientist Maurice Lemoigne in *Bacillus megaterium* in 1926 and it was a short chain length PHA, poly(3-hydroxybutyrate): P(3HB), and it was a well-studied PHA.

PHAs act as a sink for carbon and energy storage materials for redundant reducing power under limiting nutrients in the presence of excess carbon source (Brandl et al., 1988; Lageveen, 1988; Huisman et al., 1989). Prokaryotic organisms are known to produce PHB amounting to as much as 80% of their cellular dry weight (CDW). In 1983, Mcl-PHA was discovered in *Pseudomonas oleovorans* cultured in octane (De Smet et al., 1983). Archaea and eubacteria genera have so far been reported to accumulate different PHA (Zinn et al., 2001). A wide variety of PHA copolymers have been isolated from bacteria, like many cyanobacteria found in different environmental samples, including marine sediments (Wang and Bakken, 1998).

On the basis of culture condition of PHA production, bacteria are mainly divided into two major groups (Peters and Rehm, 2005). First group can synthesize PHA with the limitation of an essential nutrient as well as excess source of carbon after the growth phase such as *Ralstonia eutropha, Protomonas extorquens*, and *Protomonas oleovoran*s. The second group can synthesize PHA during the growth phase and does not require nutrient limitation like *Alcaligenes latus*, a mutant strain of *Azotobacter vinelandii* and recombinant *Escherichia coli*. These characteristics are regarded as important while production of PHA. The molecular structure of the PHA is directly depending upon the used organism, culture conditions and carbon substrate used for the organism's growth (Lebo, Gargulak, and McNally, 2001). New PHA synthesizing bacteria like *Axobacter*, *Azospirillum*, *Beggiatoa*, *Beijerinckia*, *Caulobacter*, *Clostridium*, *Halobacterium*, *Leptothrix*, *Methylocystis*, *Rhizobium*, *Thiocaspa*, *Rhodospirillum*, *Cyanobacteria*, *Paracoccus denitrificans*, *Pseudomonas stutzeri*, *P. aeruginosa*, recombinant *Bacillus subtilis*, *Delftia acidovorans*, *Azotobacter chroococcum*, *A. vinelandii*, *Comamonas* sp., *Streptomyces aureofaciens* and *Aeromonas hydrophila* are known today.

However, for the industrial production of PHA, recombinant *E. coli* with all functional genes have been introduced. These bacteria can use a wider range of cheap carbon sources, and it is relatively easier to extract and purify the polymer from these bacteria. Many new recombinant organisms like different microbes, yeasts, and plants have been introduced for PHA synthesis from cheap carbon sources like molasses, sucrose, lactose, glycerol, oils, and methane.

4.2 SUBSTRATES USED FOR PHA PRODUCTION

Since PHAs play an important role as a substitute for the petroleum-derived polymer material and its biodegradability towards environment; various factors need to be addressed to produce PHAs at the industrial scale. However, the major problem is a high cost in its commercial production and application that makes PHA production more expensive than the petroleum-derived polymers such as polyethylene and polypropylene. The ability of the bacteria to utilize cheap carbon sources is important because the significant factor for the production cost of PHA is the cost of carbon substrates (Choi and Lee, 1999). Therefore, PHA production is mainly focused on the renewable resources where bacteria can utilize waste generated from food, agro-horticultural, industrial processes, and carbon sources.

Efforts are made to develop recombinant bacterial strains able to use cheap carbon sources and effective fermentation strategies are developed and optimized to improve productivity. Some workers have tried alternative carbon sources for the production of PHAs economical. Crude carbon substrates such as corn oil (Shang et al., 2008); palm oil (Lee, 1996a); plant oil (Lee, 1996b; Khanna and Srivastava, 2005; Zinn and Hany, 2005); molasses (Byrom, 1994; Yu, 2001; Du and Yu, 2002), starch (De Smet et al., 1983; Otari and Ghosh, 2009), whey (Khanna and Srivastava, 2005) and industrial wastes (Tsuge, 2002; Shang, 2008; Hassan, 2002) are excellent substrates for the growth and polymer production and are used by different bacteria. Utilization of these plant derivatives are an indirect way of using atmospheric carbon dioxide as the carbon source for PHAs production (Byrom, 1987). However, the polymer concentration was lower than those obtained using purified carbon substrates. Therefore, there is a need to develop more efficient fermentation strategies using cheap carbon source. As compared to other processes, the use of mixed cultures has gained attention o be effective, cheaper, and potential (Bhubalan et al., 2008). In most cases of mixed cultures, the organic acids contained in wastewater or transformed from other industrial wastes were used as carbon source for the production of PHAs. Consequently, mixed culture procedures do not only help to solve the environmental pollution problem caused by organic wastes, but also help them to convert into useful materials, which is important for sustainable development and environmental protection.

4.3 STRUCTURE AND CLASSIFICATION OF PHA

The structure of PHA granules is not yet fully elucidated, but the major constituents with small amounts of proteins and lipids are known. PHA synthesis occurs intracellularly in multiple cytoplasmic inclusion bodies that are generally 200–500 nm in size and surrounded by a membrane to which proteins are bound (Sudesh et al., 2000). The hydrophobic polyester core is largely amorphous with water acting as a plasticizer prevents the crystallization. The initial investigations on PHA granules from bacteria were performed by Lee et al. (2008). PHA granules may be surrounded by a phospholipid monolayer that contains specific granule associated proteins (Figure 5.1) such as (i) PHA synthase (PhaC) (ii) PHA depolymerases and 3HB-oligomer hydroxylase (PhaZi) and (iii) phasins (PhaPs) are thought to be the major membrane-bound structural protein of the PHA-containing inclusion bodies and (iv) the regulator of phasin expression (PhaR) and also cytosolic proteins that may be non-specifically attached to the granules via hydrophobic interactions.

The polyester core is known to be surrounded by a 4 nm boundary layer, which most likely comprises of a phospholipid monolayer (Pozo et al., 2002) with embedded and attached proteins. Recently atomic force microscopy (AFM) studies on PHB granules in *Comamonas acidovoran* and *R. eutropha* (Dionisi et al., 2005; Kahar et al., 2004) confirmed the existence of an envelope surrounding the PHB granules. PHB inclusions from sonicated cells of *R. eutropha* revealed on one hand a rough and ovoid and, on the other hand, a smooth and spherical surface structure and shape. Furthermore, splits and fissures are identified at the surface of rough PHB granules. Measurements of the splits indicated a thickness of the boundary layer of 4 nm. With electron microscopy, Lundgren et al. (1964) showed that the surface of PHA granules in *B. megaterium* and *B. cereus* is covered by a membrane of about 1,520 nm thickness. AFM analysis has shown that PHA granules have an additional network layer with globular areas with 35 nm in diameter, most likely also incorporating structural phasin proteins. This was also used to show porin-like structures in the surrounding membrane, which have suggested to provide a portal to the amorphous polymer core and be the site of PHA metabolism and depolymerization. When the layer is damaged during the isolation of the granules, the crystallinity increases by 5,060%. Two physical states of the polyester occur in PHB granules: intracellular native, amorphous granules and partially crystalline granules (Solaiman et al., 2006).

4.3.1 PHA SYNTHASE (PHAC)

In *Ralstonia eutropha*, PhaC is a soluble protein in the cytoplasm, which becomes insoluble with granule binding following the onset of PHA accumulation under the cultivation conditions permissive for PHA accumulation. This is due to the PHA chain, which remains covalently linked to the enzyme until synthesis of the polymer has finished (Koller et al., 2005; Chen et al., 2006). However, the PHA synthase becomes the only PHB-granule-bound during PHB biosynthesis, when the growing hydrophobic polyester molecules are covalently linked to the enzymes during polymerization; confer amphiphilicity to the enzyme-polyester complex. According to the substrate specificities and sequence homologies, four different classes of PHA synthases were distinguished:

i. **Class I:** PHA synthases synthesize PHAs of hydroxyalkanoates of short-chain-length (PHA-Scl).
ii. **Class II:** PHA synthases prefer coenzyme A thioesters of hydroxyalkanoates of medium-chain-length (PHA-Mcl) comprising 6–14 carbon atoms as substrates and occur in strains belonging to the genus *Pseudomonas* of the rRNA homology group-I. However, in contrast to class I and class II PHA synthases, these enzymes are composed of two different subunits designated as PhaE and PhaC.
iii. **Class III:** This class synthase investigated in detail is the enzyme of *A. vinosum*. They exhibit substrate specificity with PHA-Scl.
iv. **Class IV:** PHA synthases are also composed of two types of subunits and occur in bacterial species belonging to the *Bacillus* genus (McCool and Cannon, 2001). Although the PhaC protein of class IV PHA synthases revealed high homologies to PhaC of class III PHA synthases, an activator, PhaR, is required for PHA synthase activity.

The class I and class III PHA synthases of a-proteobacterium *R. eutropha* and ç-proteobacterium *A. vinosum*, respectively, are regarded as model enzymes for studying PHA-Scl biosynthesis in bacteria, whereas the PHA-Mcl synthases of *P. oleovorans* and *P. putida* represent the most detailed studied class II PHA synthases. PHA synthases are also responsible for the MW of PHA synthesized. Metabolic factors are the intracellular concentration of the PHA synthases, the enzyme-to-substrate ratio, and probably also enzymes capable of hydrolyzing PHAs, such as intracellular PHA depolymerases or esterases and lipases (Jaeger et al., 1995). The absence of PHA depolymerases results in the formation of PHAs exhibiting a higher molecular mass as in the case of recombinant strains of *E. coli* expressing the PHA operon cloned from *R. eutropha* (Kusaka et al., 1997). The level of PhaC expression has a significant influence role in the molecular mass of polyester. A higher concentration of active PHA synthase protein resulted in a lower molecular mass of the resulting PHA molecules (Sim et al., 1997).

4.3.2 PHA DEPOLYMERASE (PHAZI)

Several bacteria express extracellular depolymerases, which are secreted into the environment to degrade PHA released from dead bacteria. The immature extracellular depolymerases (Pha Ze) contain in general N-terminal signal peptides of 25–58 amino acids, a large catalytic domain at the N-terminal region, a substrate binding domain localized at the C-terminal region, and a linking domain connecting the catalytic domain with the substrate-binding domain (Jendrossek, 2002; Yong, 2008). In contrast to the well-studied extracellular PHA depolymerases, intracellular PHA depolymerases (PhaZi) have been far less investigated, although they play an important role for the overall PHA metabolism. In *B. megaterium*, *Zooglea ramigera*, *Sinorhizobium melioti*, and *R. eutropha* PHA hydrolyzing activity has been demonstrated. Very recently, four additional intracellular PHA depolymerases were identified in *R. eutropha* by York et al. (2003), PhaZ2 and PhaZ3, PhaZ4, and PhaZ5.

4.3.3 PHASINS (PHAP1)

The structural protein phasins (PhaP1) represent a class of probable noncatalytic proteins consisting of a hydrophobic domain, which associates with the surface of the PHB granules, and of a predominantly hydrophilic/amphiphilic domain exposed to the cytoplasm of the cell. PhaP1 adheres very tightly in vivo to native as well as in vitro to artificial PHB granules. PhaP1 is synthesized in large quantities under storage conditions and represents as much as 5% of the total protein (Tian et al., 2005). By this, PhaP1 protects the host cell by covering the hydrophobic surface of the polymer thereby preventing other proteins binding to the hydrophobic granules (Griebel and Merrick, 1971). The anchoring region of the PHA-Scl is located at the carboxy terminus of the accumulating bacteria *Rhodococcusr uber* at the carboxy terminus and it is found that the phasin molecules truncated at the carboxy-terminal region preventing their capability to bind to PHA granules (Pieper-Furst et al., 1995). The addition of phasins accelerated the PHA synthesis rate and regulates the size of PHA inclusions. Overexpression of PhaP, on the other hand, results in the formation of many small granules in the cell. The addition of PhaP isolated from *R. eutropha* increased the activity of class II PHA synthase from *P. aeruginosa* by approximately 50%. Reverse transcription and polymerase chain reaction (RT-PCR) analysis clearly demonstrated that PhaP2, PhaP3, and PhaP4 were also transcribed under conditions permissive for PHB biosynthesis and accumulation (Mayer et al., 1996).

4.3.4 TRANSCRIPTIONAL REPRESSOR (PHAR)

Transcriptional repressor PhaR, the regulator of phasin expression was first detected in *R. eutropha*. PhaR is proposed to be a repressor protein of transcription that binds to the PhaP1 upstream region in *R. eutropha*, thereby repressing expression of PhaP1. Biosynthesis and accumulation of PHB derepresses PhaR. PhaR has the capability to bind to at least three different targets in cells of *R. eutropha*, (i) promoter region of PhaP1, (ii) promoter region of PhaR, and (iii) surface of PHB granules. PhaR can work on five different situations to regulate PhaP:

1. If cells are cultivated under conditions not permissive to PHB biosynthesis, PhaR cannot bind to PHB granules because they do not exist in the cells. The cytoplasmic concentration of PhaR is sufficiently high to repress transcription of PhaP1 and no PhaP1 protein is formed and detectable in the cytoplasm.

2. In suitable conditions, biosynthesis of PHB starts with constitutively expression of PHA synthase (PhaC) which starts to synthesize PHB molecules that remain covalently linked to the enzyme. At the beginning, small micelles are formed which gradually become larger and constitute the nascent PHB granules. PhaC no longer covers the PHB granule surface entirely, and proteins with a binding to the hydrophobic surface like PhaR to the granules. This lowers the cytoplasmic concentration of PhaR.
3. From a certain point the cytoplasmic concentration of PhaR becomes so low that it can no longer repress transcription of PhaP1.
4. The concentration of soluble PhaP1 in the cytoplasm remains beyond a detectable level. The PHB granules grow and reach their maximum size; PhaP protein is being continuously synthesized in sufficient amounts.
5. When the PHB granules have reached the maximum possible size according to the physiological conditions almost the entire surface will be covered by PhaP1 protein, and the latter is displacing PhaR protein from the PHB granules. Consequently, the cytoplasmic concentration of PhaR increases and it will exceed the threshold concentration required to repress again transcription of PhaP1. Therefore, PhaP1 protein is no longer synthesized. This mode of regulation ensures that PhaP1 is not produced in higher amounts than required to cover the surface of PHB granules. In addition, the binding capacity of PhaR to the promoter region prevents over expression of this repressor protein that is, therefore, under auto-control.

KEYWORDS

- ***Alcaligenes latus***
- ***Escherichia coli***
- **micelles**
- **poly(3-hydroxybutyrate)**
- **polyhydroxyalkanoates**
- **polyhydroxybutyrate**

REFERENCES

Albuquerque, M. G. E., et al., (2007). Strategies for the development of a side stream process for polyhydroxyalkanoate (PHA) production from sugar cane molasses. *J. Biotechnol., 130*, 411–421.

Anderson, A. J., & Dawes, E. A., (1990). Occurrence, metabolism, metabolic role, and industrial use of bacterial polyhydroxyalkanoates. *Microbiol. Rev., 54*, 450–472.

Barnard, G. N., & Sanders, J. K., (1989). The poly-beta-hydroxybutyrate granule *in vivo*. A new insight based on NMR spectroscopy of whole cells. *J. Biol. Chem., 264*, 3286–3291.

Berlanga, M., et al., (2006). Rapid spectrofluorometric screening of poly-hydroxyalkanoate-producing bacteria from microbial mats. *International Microbiology, 9*, 95–102.

Bhubalan, K., et al., (2008). Controlled biosynthesis and characterization of poly (3-hydroxybutyrate-co-3-hydroxyvalerate-co-3-hydroxyhexanoate) from mixtures of palm kernel oil and 3HV-precursors. *Polym. Degrad. Stab., 93*, 17–23.

Brandl, H., et al., (1988). *Pseudomonas oleovorans* as a source of poly (β-hydroxyalkanoates) for potential applications as biodegradable polyesters. *Appl. Environ. Microbiol., 54*, 1977–1982.

Braunegg, G., et al., (1998). Polyhydroxyalkanoates, biopolyesters from renewable resources: Physiological and engineering aspects. *J. Biotechnol., 65*, 127–161.

Brown, Jr. R. M., (1996). The biosynthesis of cellulose. *J. Macromol Sci. Pure Appl. Chem., 33*, 1345–1373.

Byrom, D., (1987). Polymer synthesis by microorganisms: Technology and economics. *Trends Biotechnol., 5*, 246–250.

Byrom, D., (1994). Polyhydroxyalkanoates, In: Mobley, D. P., (ed.), *Plastic from Microbes: Microbial Synthesis of Polymers and Polymer Precursors* (pp. 5–33). Hanser Munich.

Chen, C. W., et al., (2006). Enzymatic extruded starch as a carbon source for the production of poly (3-hydroxybutyrate-co-3-hydroxyvalerate) by *Haloferax mediterranei*. *Proc Biochem., 41*, 2289–2296.

Chen, G. Q., & Wu, Q., (2005a). The application of polyhydroxyalkanoates as tissue engineering materials. *Biomaterials, 26*, 6565–6578.

Chen, G. Q., & Wu, Q., (2005b). Microbial production and applications of chiral hydroxyalkanoates. *Appl. Microbiol. Biotechnol., 67*, 592–599.

Chen, G., et al., (2001). Industrial scale production of poly (3-hydroxybutyrate-co-3-hydroxyhexanoate). *Appl. Microbiol. Biotechnol., 57*, 50–55.

Choi, J., & Lee, S. Y., (1999). Factors affecting the economics of polyhydroxyalkanoate production by bacterial fermentation. *Appl. Microbiol. Biotechnol., 51*, 13–21.

Chowdhury, A. A., (1963). Polyhydroxybutyric acid degrading bacteria and exo-enzymes. *Arch. Microbiol., 47*, 167–200.

De Smet, M. J., et al., (1983). Characterization of intracellular inclusions formed by *Pseudomonas oleovorans* during growth on octane. *J. Bacteriol., 154*, 870–878.

Dionisi, D., et al., (2005). Olive oil mill effluents as a feedstock for the production of biodegradable polymers. *Water Res., 39*, 2076–2084.

Du, G., & Yu, J., (2002). Green technology for conversion of food scraps to biodegradable thermoplastic polyhydroxyalkanoates. *Environ. Sci. Technol., 36*, 5511–5516.

Flieger, M., et al., (2003). Biodegradable plastics from renewable sources. *Folia Microbiol., 48*, 2–44.

Griebel, R. J., & Merrick, J. M., (1971). Metabolism of poly-hydroxybutyrate: Effect of mild alkaline extraction on native poly-hydroxybutyrate granules. *J. Bacteriol., 108*, 782–789.

Hassan, M. A., et al., (2002). A proposal for zero-emission from palm oil industry incorporating the production of polyhydroxyalkanoates from palm oil mill effluent. *J. Chem. Eng., 35*, 9–14.

Huisman, G. W., et al., (1989). Synthesis of poly-3-hydroxyalkanoates is a common feature of fluorescent *Pseudomonads*. *Appl. Environ. Microbiol., 55*, 1949–1954.

Jaeger, K. E., et al., (1995). Substrate specificities of bacterial polyhydroxyalkanoate depolymerases and lipases: Bacterial lipases hydrolyze poly (ω-hydroxyalkanoates). *Appl. Environ. Microbiol., 61*, 3113–3118.

Kadouri, D., et al., (2005). Ecological and agricultural significance of bacterial polyhydroxyalkanoates. *Crit. Rev. Anal. Chem., 31*, 55–67.

Kahar, P., et al., (2004). High yield production of polyhydroxyalkanoates from soybean oil by *Ralstoniaeutropha* and its recombinant strain. *Polym. Degrad. Stab., 83*, 79–86.

Khanna, S., & Srivastava, A. K., (2005). Recent advances in microbial polyhydroxyalkanoates. *Process Biochem., 40*, 607–619.

Koller, M., et al., (2005). Production of polyhydroxyalkanoates from agricultural waste and surplus materials. *Biomacromolecules, 6*, 561–565.

Koller, M., et al., (2008). Polyhydroxyalkanoate production from whey by *Pseudomonas hydrogenovora*. *Bioresour. Technol., 99*, 4854–4863.

Kusaka, S., et al., (1997). Molecular mass of poly [(R)-3-hydroxybutyric acid] produced in a recombinant *Escherichia coli*. *Appl. Microbiol. Biotechnol., 47*, 140–143.

Lageveen, R. G., et al., (1989). Formation of polyesters by *Pseudomonas oleovorans*: Effect of substrates on formation and composition of poly-(R)-3-hydroxyalkanoates and poly-(R)-3-hydroxyalkenoates. *Appl. Environ. Microbiol., 54*, 2924–2932.

Lebo, Jr. S. E., Gargulak, J. D., & McNally, T. J., (2001). *Lignin Kirk–Othmer Encyclopedia of Chemical Technology*. John Wiley and Sons Inc.

Lee, S. Y., (1995). Review bacterial polyhydroxyalkanoates. *Biotechnol. Bioeng., 49*, 1–14.

Lee, S. Y., (1996a). Bacterial polyhydroxyalkanoates. *Biotechnol. Bioeng., 49*, 1–14.

Lee, S. Y., (1996b). Plastic bacteria? Progress and prospects for polyhydroxyalkanoate production in bacteria. *Trends Biotechnol., 14*, 431–438.

Lee, W. H., et al., (2008). Biosynthesis of polyhydroxyalkanoate copolymers from mixtures of plant oils and 3-hydroxyvalerate precursors. *Bioresour Technol., 99*, 6844–6851.

Lemoigne, M., (1926). Produits de deshydrationet de polymerisation de l'acide b-oxybutyric. *Bull. Soc. Chem. Biol., 8*, 770–782.

Lundgren, D. G., et al., (1964). Characterization of poly-β-hydroxybutyrate extracted from different bacteria. *J. Bacteriol., 89*, 245–251.

Madison, L. L., & Huisman, G. W., (1999). Metabolic engineering of poly (3-hydroxyalkanoates): From DNA to plastic. *Microbiol. Mol. Biol. Rev., 63*, 21–53.

Mayer, F., et al., (1996). Electron microscopic observation on the macromolecular organization of the boundary layer of bacterial PHA inclusion bodies. *J. Gen. Appl. Microbiol., 42*, 445–455.

Mc Cool, G. J., & Cannon, M. C., (2001). PhaC and Pha R are required for polyhydroxyalkanoic acid synthase activity in *Bacillus megaterium*. *J. Bacteriol., 183*, 4235–4243.

Mergaert, J., et al., (1993). Microbial degradation of poly (3-hydroxybutyrate) and poly(3-hydroxybutyrate-co-3-hydroxyvalerate) in soils. *Appl. Environ. Microbiol., 59*, 3233–3238.

Nielsen, L., (2007). Polyhydroxyalkanoate production in sugarcane recognizing temporospatial complexity. *J. Biotechnol., 131*, S28, S29.

Oeding, V., & Schlegel, H. G., (1973). Beta-ketothiolase from *Hydrogenomonas eutropha* HI6 and its significance in the regulation of poly-beta-hydroxybutyrate metabolism. *Biochem. J., 134*, 239–248.

Otari, S. V., & Ghosh, S. J., (2009). Production and characterization of the polymer polyhydroxybutyrate-co-polyhydroxyvalerate by *Bacillus megaterium* NCIM 2475. *Curr Res. J. Biol. Sci., 1*, 23–26.

Peters, V., & Rehm, B. H. A., (2005). *In vivo* monitoring of PHA granule formation using GFP-labelled PHA synthases. *FEMS Microbiol. Lett., 248*, 93–100.

Philip, S., Keshavarz, T., & Roy, I., (2007). Polyhydroxyalkanoates: Biodegradable polymers with a range of applications. *J. Chem. Technol. Biotechnol., 82*, 233–247.

Pieper-Furst, U., et al., (1995). Identification of the region of a 14-kilodalton protein of *Rhodococcus ruber* that is responsible for the binding of this phasin to polyhydroxyalkanoic acid granules. *J. Bacteriol., 177*, 2513–2523.

Poirier, Y., et al., (1995). Production of polyhydroxyalkanoates, a family of Biodegradable plastics and elastomers, in bacterial and plant. *Biotechnol., 13*, 142–150.

Pozo, C., et al., (2002). Effects of culture conditions on the production of polyhydroxyalkanoates by *Azotobacter chroococcum* H23 in media containing a high concentration of alpechin (wastewater from olive oil mills) as primary carbon source. *J. Biotechnol., 97*, 125–131.

Shang, L., et al., (2008). Mass production of medium-chain-length poly (3-hydroxyalkanoates) from hydrolyzed corn oil by fed-batch culture of *Pseudomonas putida*. *World J. Microbiol. Biotechno., 24*, 2783–2787.

Shishatskaya, E. I., et al., (2004). Tissue response to the implantation of biodegradable polyhydroxyalkanoate sutures. *J. Mater. Sci. Mater. Med., 15*, 719–728.

Sim, S. J., et al., (1997). PHA synthase activity controls the molecular weight and polydispersity of polyhydroxybutyrate *in vivo*. *Nat. Biotechnol., 15*, 63–67.

Solaiman, D., et al., (2006). Biosynthesis of medium-chain-length poly (hydroxyalkanoates) from soy molasses. *Biotechnol. Lett., 28*, 57–162.

Steinbuchel, A., (1991). In: Byrom, D., (ed.), *Biomaterials: Novel Materials from Biological Sources* (pp. 123–213). Macmillan., New York, NY.

Steinbuchel, A., (2001). Perspectives for biotechnological production and utilization of biopolymers: Metabolic engineering of polyhydroxyalkanoate biosynthesis pathways as a successful example. *Macromol. Biosci., 1*, 1–24.

Steinbuchel, A., (2005). Non-biodegradable biopolymers from renewable resources: Perspectives and impacts. *Curr. Opin. Biotechnol., 16*, 607–613.

Steinbuchel, A., et al., (1995). Considerations on the structure and biochemistry of bacterial polyhydroxyalkanoic acid inclusions. *Can. J. Microbiol., 41*, 94–105.

Sudesh, K., et al., (2000). Synthesis, structure, and properties of polyhydroxyalkanoates: Biological polyesters. *Prog. Polym. Sci., 25*, 1503–1555.

Taji, H. (2002). Polylactides. In: Doi, Y., & Steinbuchel, A., (eds.), *Polyesters III: Applications and Commercial Products* (pp. 129–177). Weinheim, Germany: Wiley-VCH.

Tian, J., et al., (2005). Analysis of transient polyhydroxybutyrate production in *Wautersia eutropha* H16 by quantitative Western analysis and transmission electron microscopy. *J. Bacteriol., 187*, 3825–3832.

Tsuge, T., (2002). Metabolic improvements and use of inexpensive carbon sources in microbial production of polyhydroxyalkanoates. *J. Biosci. Bioeng., 94*, 579–584.

Wang, J. G., & Bakken, L. R., (1998). Screening of soil bacteria for poly-beta-hydroxybutyric acid production and its role in the survival of starvation. *Microb. Ecol., 35*, 94–101.

Williams, S. F., et al., (1999). PHA applications, addressing the price performance issue: I. Tissue engineering. *Int. J. Biol. Macromol., 25*, 111–121.

Williamson, D. H., & Wilkinson, J. F., (1958). The isolation and estimation of poly-β-hydroxybutyrate inclusions of *Bacillus* species. *J. Gen. Microbiol., 19*, 198–209.

York, G. M., et al., (2003). *Ralstoniaeutropha* H16 encodes two and possibly three intracellular poly [D-(-)-3-hydroxybutyrate] depolymerase genes. *J. Bacteriol., 185*, 3788–3794.

Yu, J., (2001). Production of PHA from starch wastewater via organic acids. *J. Biotechnol., 86*, 105–112.

Zinn, M., & Hany, R., (2005). Tailored material properties of polyhydroxyalkanoates through biosynthesis and chemical modification. *Advanced Engg. Materials, 7*, 408–411.

CHAPTER 5

Classification of PHAs

Depending on the substrate provided, many microorganisms can include a wide variety of 3-hydroxy fatty acids in the PHA. More than 100 different monomer units have been identified as constituents of PHA in >300 different microorganisms (Jendrossek and Handrick, 2002).

Depending on the number of carbon atoms in the monomeric unit, PHAs are classified into two different types:

i. Short chain length PHAs (scl-PHAs; 3–5 C-atoms); and
ii. Medium chain length PHAs (Mcl-PHA; 6 or more C-atoms).

Also, depending on the kind of monomer present, PHAs can be a homopolymer containing only one type of hydroxyalkanoate as the monomer unit, e.g., P(3HB), P(3HHx) or a heteropolymer containing more than one kind of hydroxyalkanoate as monomer units, e.g., poly(3-hydroxybutyrate-co-3-hydroxyvalerate), P(3HB-co-3HV), poly(3-hydroxyhexa-noate-co-3-hydroxyoctanoate), P(3HHx-co-3HO), poly-(3-hydroxybutyrate-co-3-hydroxyhexanoate), P(3HB-co-3HHx) (Gerngross et al., 1993; Liebergesell et al., 1994). Due to the continuous discovery of new homopolymers and copolymers, the polymers of the PHA family are constantly increasing in number. Moreover, saturated, unsaturated, halogenated, branched, and aromatic side chains in (R)–3HA monomeric units have also been found in the sequence of microbial PHA (McCool and Cannon, 2001; Mukai et al., 1993). This is in turn resulting in the availability of PHAs with a wide range of chemical structures and assortment of properties. The properties of PHAs vary significantly depending on their monomer content and consequently can be customized

Bacterial Biopolymers. B. K. Konwar, PhD (Author)

by controlling their compositions. The differences in their properties have also been shown to greatly affect the mode/rate of degradation in aqueous or biological media.

5.1 METABOLIC PATHWAY OF PHA BIOSYNTHESIS

Microorganisms can produce intracellular PHA inclusions in their stationary growth phase under nutritional unbalanced growth conditions but an excess of carbon source. Different types of enzymes are involved in running the metabolic process of the bacteria for the formation of the inclusion bodies. Based on the types of monomers incorporated into PHA, various metabolic pathways have been shown to be involved in the PHA biosynthesis. The schematic illustration of the metabolic pathway of PHA biosynthesis is shown in Figure 5.1 in broad-spectrum.

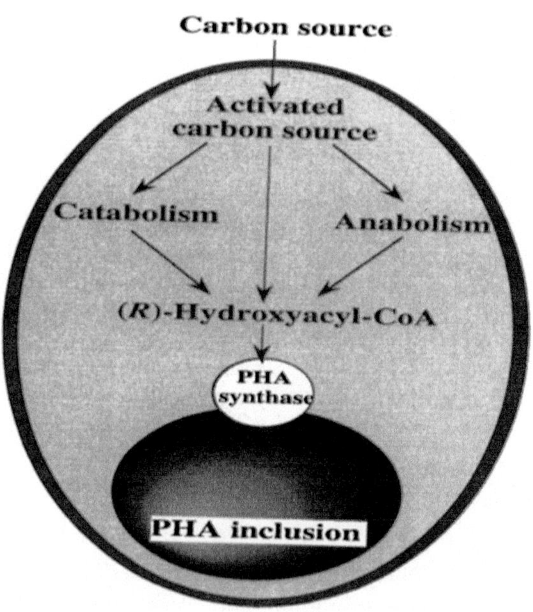

FIGURE 5.1 General scheme for the metabolic pathways of PHA synthesis from different carbon sources within a bacterial cell.

Source: Reprinted with permission from Sudesh et al., 2000. © Elsevier.

5.2 ENZYMES INVOLVED

Depending on functions, the main key enzymes involved in the PHA biosynthesis are categorized into three. The key enzyme in PHA biosynthesis is PHA synthase (PhaC) which mainly polymerize the monomeric groups of carbon source intake by the microorganisms. The second type of enzyme is β-ketothiolase (PhaA) which is involved in the intermediate metabolic pathway program. The third type of enzyme is acetoacetyl CoA reductase (PhaB) which basically reflects the main role in oxidation-reduction reaction in PHA metabolism process. Moreover, the genes and enzymes involved in PHAs synthesis have evolved features to different microbial groups.

5.2.1 PHA SYNTHASE (PHAC)

This enzyme is mainly associated with the polymerization reaction from hydroxyalkanoic acids as substrates. The substrate specificity of PHA synthases may determine the ability of microorganisms to synthesize a form of PHA. PHA synthases may be divided into four classes based on their primary structure, the subunit composition, and the substrate specificity (Kusaka et al., 1997; Sim et al., 1997) (Figure 5.2):

i. **Class I:** PHA synthases consists of only one type of subunit (PhaC). Its molecular weight (MW) ranges between 61 and 73 kDa and catalyzes polymerization of 3–5 carbon atoms and mainly utilizes CoA thioesters of 3-HAs, 4-HAs, and 5-Has (Kraak et al., 1997).
ii. **Class II:** This type of PHA synthases is encoded by two different genes, PhaC1 and PhaC2, again 61 to 73 kDa in size, each catalyzes the polymerization of CoA thioesters of various $3HA_{Mcl}$ comprising 6–14 carbon atoms (Kusaka et al., 1997). Two PhaC genes are separated by the phaZ gene, encoding the PHA depolymerase.
iii. **Class III:** This class synthases are composed of two subunits (PhaC and PhaE) with MWs 40 kDa each and possess substrate specificities like class I type. These PHA synthases prefer coenzyme A thioester of $3HA_{Scl}$ with 6–8 carbon atoms (Jendrossek, 2002).

iv. **Class IV:** This class synthases are composed of two subunits (PhaC and PhaR) that can utilize coenzyme A thioester of R-3-hydroxy fatty acids with 3–5 carbon atoms (Haas et al., 2008). The MWs of the two subunits are 40 kDa and 22 kDa respectively.

5.2.2 B-KETOTHIOLASE (PHAA)

This enzyme is mainly responsible for the degradation of fatty acids by thiolytic cleavage in PHA biosynthesis process. Based on the substrate specificity, this enzyme is divided into two groups (Chowdhury, 1963). The first group of β-ketothiolase has broad substrate specificity with C_4–C_{16} and mainly involved in the degradation of fatty acids. The second group of β-ketothiolase has a narrow range of substrate specificity with C_3–C_5 and mainly responsible for the biosynthesis. Both the types of β-ketothiolase exist as homotetramer with molecular mass of 160–190 kDa.

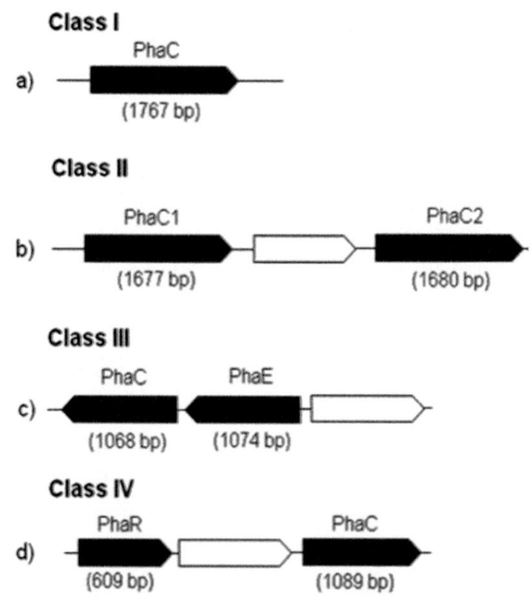

FIGURE 5.2 Schematic representation of different classes of the PHA synthases (Adapted from Rehm, 2003)].

5.2.3 ACETOACETYL COA REDUCTASE (PHAB)

The acetoacetyl CoA reductase (PhaB) is NADPH dependent enzyme and catalyzes the second step in PHB biosynthetic pathway by converting the acetoacetyl CoA into 3-hydroxybutyryl CoA through oxidation-reduction process. These are basically of two types depending on their origin from different genera: (i) NADH dependent and (ii) NADPH dependent. PhaB is a homotetramer with identical subunits with molecular masses ranging from 85–140 kDa.

5.3 PHA BIOSYNTHESIS PATHWAY

Bacteria can undergo PHA biosynthesis through three different biosynthetic pathways. The precursor carbon source has been provided the three types of pathways which were summarized in a single schematic diagram (Figure 5.3).

5.3.1 PATHWAY I

Pathway I is related to the formation of scl PHA molecules and the precursor molecule is acetyl CoA provided by glycolysis process within the bacterial cell. In the case of *R. eutropha* pathway I is generally observed. This pathway consists of 3 enzymatic reactions catalyzed by three different enzymes (Figure 5.4, Pathway I). Two acetyl-CoA moieties are condensed to acetoacetyl-CoA by a b-ketothiolase (PhaA). The product then undergoes reduction by an NADPH-dependent reductase (PhaB) which produces (R)-isomer of 3-hydroxybutyryl-CoA (25). On the other hand, in *Rhodospirillum rubrum* which shares almost similar PHA biosynthesis pathway as *R. eutropha*, the reductase which is an NADH-dependent isoenzyme, gives rise to (S)-isomer of 3-hydroxybutyryl-CoA. Lastly, (R)-3-hydroxybutyryl-CoA monomers are polymerized by PHA synthase (PhaC) to P(3HB).

5.3.2 PATHWAY II

Another type of PHA biosynthesis pathway is mainly involved in the PHA_{Mcl} synthesis through fatty acid β-oxidation pathway (Figure 5). This pathway is exhibited by the *Pseudomonads* belonging to rRNA-homology-group I which derives the 3-hydroxyacyl-CoA substrates of C_6–C_{14} for

PHA_{Mcl} synthase to synthesize PHA_{Mcl} from various alkanes, alkanols or alkanoates (Ayub et al., 2004). Here, the fatty acids are first converted to the corresponding acyl-CoA thioesters which are then oxidized by fatty acid ß-oxidation via trans-2-enoyl-CoA and (S)-3-hydroxyacyl-CoA to form 3-ketoacyl-CoA. 3-ketoacyl-CoA is then cleaved by a ß-ketothiolase to form acetyl-CoA and an acyl-CoA comprising of two less carbon atoms

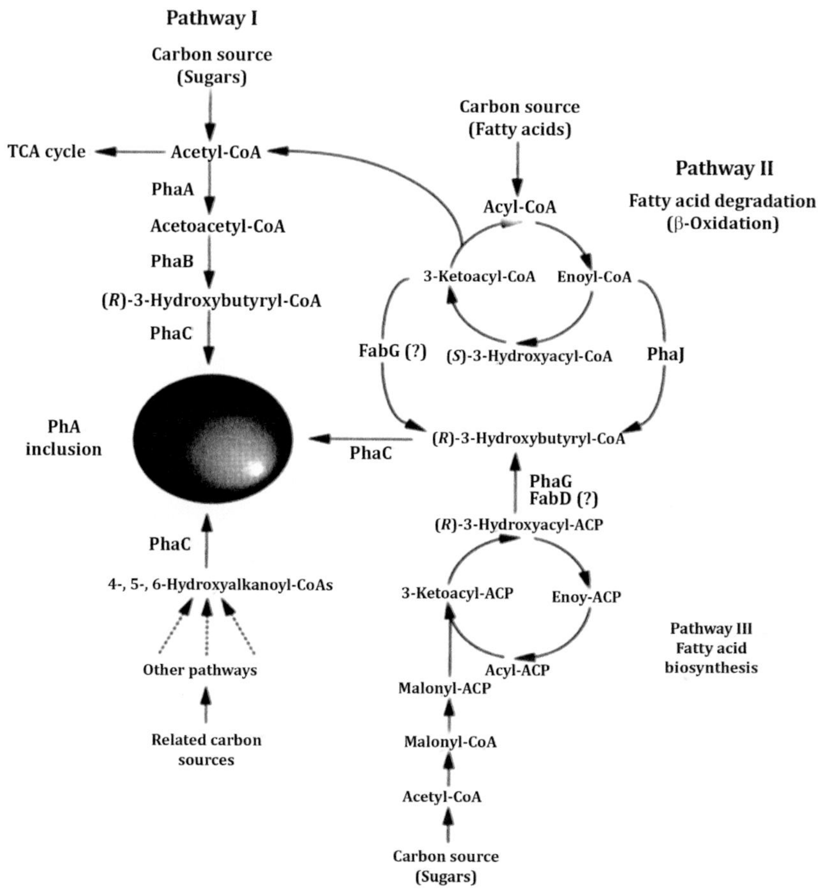

FIGURE 5.3 Cyclic metabolic nature of P(3HB) biosynthesis and degradation in bacteria. PhaA, b-ketothiolase; PhaB, NADPH dependent acetoacetyl-CoA reductase; PhaC, PHA synthase; PhaZ, PHA depolymerase; 1, dimer hydrolase; 2, (R)-3-hydroxybutyrate dehydrogenase; 3, acetoacetyl-CoA synthetase; 4, NADH-dependent acetoacetyl-CoA reductase.

Source: Reprinted with permission from Sudesh et al., 2000. © Elsevier.

as compared to the acyl-CoA that entered the first cycle. Specific enzymes such as the enoyl-CoA hydratase (PhaJ) and 3-ketoacyl-CoA reductase (FabG) are apparently involved in the conversion of fatty acid β-oxidation which intermediates into suitable monomers that can be polymerized by the PHA synthase to synthesis PHA_{Mcl}. This pathway is mainly referred to as the *P. oleovorans* PHA biosynthetic pathway.

5.3.3 PATHWAY III

The third type of PHA biosynthesis pathway occurs in most *Pseudomonads* belonging to rRNA homology group I except *P. oleovorans* which can also synthesize PHA_{Mcl} from structurally unrelated, simple, and inexpensive carbon sources such as glucose, sucrose, and fructose. Therefore, this pathway is also referred to as *P. aeruginosa* PHA biosynthetic pathway. In this case, 3-hydroxyacyl monomers are derived from the *de novo* fatty acid biosynthesis pathway. Since the fatty acid biosynthesis intermediate is in the form of (R)-3-hydroxyacyl-ACP, an additional biosynthetic step is needed to convert it into (R)-3-hydroxyacyl-CoA form. Therefore, an enzyme 3-hydroxyacyl-CoA-ACP transferase (PhaG) is involved in channeling the intermediates of the *de novo* fatty acid biosynthesis pathway to PHA biosynthesis (Hoffmann et al., 2000).

KEYWORDS

- **acetoacetyl**
- **hydroxyalkanoate**
- **microorganisms**
- **monomers**
- ***R. eutropha***
- ***Rhodospirillum rubrum***

REFERENCES

Ayub, N. D., et al., (2004). Polyhydroxybutyrate-producing *Pseudomonas* sp. isolated from Antarctic environments with high stress resistance. *Curr. Microbiol., 49*, 170–174.

Chowdhury, A. A., (1963). Polyhydroxybutyric acid degrading bacteria and exo-enzymes. *Arch. Microbiol., 47*, 167–200.

Gerngross, T. U., et al., (1993). Immunocytochemical analysis of poly-β-hydroxybutyrate (PHB) synthase in *Alcaligenes eutrophus* H16: Localization of the synthase enzyme at the surface of the of PHB granules. *J. Bacteriol., 175*, 5289–5293.

Haas, R., et al., (2008). Production of poly (3-hydroxybutyrate) from waste potato starch. *Biosci. Biotechnol. Biochem., 72*, 253–256.

Hoffmann, N., et al., (2000). The *Pseudomonas aeruginosa* phaGgene product is involved in the synthesis of polyhydroxyalkanoic acid consisting of medium-chain-length constituents from non-related carbon sources. *FEMS Microbiol. Lett., 184*, 253–259.

Jendrossek, D., & Handrick, R., (2002). Microbial degradation of polyhydroxyalkanoates. *Annu. Rev. Microbiol., 56*, 403–432.

Jendrossek, D., (2002). Extracellular polyhydroxyalkanoate depolymerases: The key enzymes of PHA degradation. In: Doi, Y., & Steinbuchel, A., (eds.), *Biopolymers* (pp. 41–84). Wiley-VCH: Weinheim, Germany.

Kraak, M. N., et al., (1997). Polymerase C1 levels and poly(R-3-hydroxyalkanoate) synthesis in wild-type and recombinant *Pseudomonas* strains. *J. Bacteriol., 179*, 4985–4991.

Kusaka, S., et al., (1997). Molecular mass of poly [(R)-3-hydroxybutyric acid] produced in a recombinant *Escherichia coli*. *Appl. Microbiol. Biotechnol., 47*, 140–143.

Liebergesell, M., et al., (1994). Purification and characterization of the poly (hydroxyalcanoic acid) synthase from *Chromatium vinosum* and localization of the enzyme at the surface of poly (hydroxyalkanoic acid) granules. *Eur. J. Biochem., 226*, 71–80.

McCool, G. J., & Cannon, M. C., (2001). PhaC and PhaR are required for polyhydroxyalkanoic acid synthase activity in *Bacillus megaterium*. *J. Bacteriol., 183*, 4235–4243.

Mukai, K., et al., (1993). Substrate specificities in hydrolysis of polyhydroxyalkanoates by microbial esterases. *Biotechnol. Lett., 15*, 601–604.

Poirier, Y., et al., (1995). Production of polyhydroxyalkanoates, a family of biodegradable plastics and elastomers, in bacterial and plant. *Biotechnol., 13*, 142–150.

Rehm, B. H. A., (2003). Polyester synthases: Natural catalysts for plastics, *Biochem. J., 376*, 15–33.

Sim, S. J., et al., (1997). PHA synthase activity controls the molecular weight and polydispersity of polyhydroxybutyrate *in vivo*. *Nat Biotechnol., 15*, 63–67.

Sudesh, K., et al., (2000). Synthesis, structure, and properties of polyhydroxyalkanoates: Biological polyesters. *Prog. Polym. Sci., 25*, 1503–1555.

CHAPTER 6

Detection, Extraction, and Characterization of PHAs

6.1 DETECTION

The lipophilic dyes like Sudan black B (Potter et al., 2004), Nile blue A (Ostle and Holt, 1982) and Nile red (Spiekermann et al., 1999) are used to dye and detect the hydrophobic PHAs. The lipophilic dye Sudan black B is more soluble in lipid material and imparts a black blue color to the PHA granule. Nile blue A, a water-soluble basic oxazine is better than Sudan and does not stain other inclusion bodies like glycogen and others. It has more specific affinity to stain only PHAs (Ostle and Holt, 1982) giving bright orange fluorescence at the wavelength of 460 nm. Nile red has been observed to produce strong orange fluorescence with emission maximum at 598 nm with the maximum excitation wavelength of 543 nm on binding to P(3HB) granules in cells of *Cupria vidusnecator* (Pieper-Furst et al., 1995).

6.2 EXTRACTION

The recovery process of PHA is costly; a variety of methods have been adopted for PHAs extraction from bacterial cells. For the extraction, the bacterial cell biomass is harvested from the culture broth using centrifugation, filtration, and flocculation-centrifugation. Cells are lyophilized and subjected to the extraction methods by employing solvent extraction or by digesting the non-PHA biomass (Ramsay et al., 1994).

Bacterial Biopolymers. B. K. Konwar, PhD (Author)

6.2.1 SOLVENT

The solvent extraction methods for PHAs recovery involve the use of chlorinated solvents like methylene chloride, chloroform, propylene carbonate, or dichloroethane. The dissolved polymer is filtered, concentrated, and precipitated by using chilled non-solvents such as methanol or ethanol with vigorous shaking. The drawback of the method for the commercial purpose is the requirement of large volume of solvent (Braunegg et al., 1998; Salehizadeh and Loosdrecht, 2004.), it is hazardous for the operator and environment, during extraction destroys the natural morphology of PHA granules (Asrar et al., 2002). On the contrary, the method has some advantages over the other methods. There is no degradation of the polymer and can be useful for medical application with the elimination of endotoxin produced by gram negative bacteria (Kim and Lenz, 2001).

6.3 DIGESTION OF PHA

6.3.1 BY SURFACTANTS

Surfactants, such as anionic sodium dodecyl sulfate (SDS) disrupt cells by incorporating itself into the lipid bilayer membrane of the bacteria. Higher concentration of the surfactant breaks the cellular membrane to produce surfactant micelles and membrane phospholipids, which leads to the release of P(3HB). Surfactant solubilizes not only proteins, but also other non-PHA cellular materials and thus surfactant alone cannot give a high purity PHA (>97%). Furthermore, higher dose of surfactant is not cost-effective.

6.3.2 BY SODIUM HYPOCHLORITE

Sodium hypochlorite digestion method is another recovery method for PHA extraction. Sodium hypochlorite breaks open the bacterial cells which solubilizes non PHA material and liberates the intracellular granules. The sodium hypochlorite digestion causes degradation of the polymer up to 50% in its molecular weight (MW) (Berger et al., 1989). The combination of chloroform and sodium hypochlorite can significantly reduce the MW of PHA. The hypochlorite digests cells releasing the PHA, which get dissolved in chloroform and thus gets protected from degradation (Hahn et al., 1993). But this also imparts environmental hazards and leads to a higher cost.

6.3.3 *BY ENZYMES*

Enzymes like alcalase, neutrase, lecitase, and lysozyme are used for the digestion which is followed by washing with anionic surfactant like SDS to solubilize non-P(3HB). Both solubilized and non-solubilized cell compounds are separated by centrifugation with subsequent washing with anionic surfactant and then flocculation. Enzymatic digestion is a good recovery process for PHA extraction but leads to high cost.

6.4 QUANTIFICATION AND COMPOSITIONAL ANALYSIS OF PHA

Quantification of PHA is done by gravimetric measurement from lyophilized cell biomass. The freeze-dried biomass is digested using sodium hypochlorite and the sediment material extracted by chloroform, followed by precipitation with diethyl ether methanol or acetone. Qi and Rehm (2001) has developed another method for PHA quantification. In this method PHA is converted to crotonic acid by treating with concentrated sulfuric acid and measured the product by taking optical density at 235 nm. However, this method of estimation is less accurate as because chloroform solubilize impurities of crude polymer which absorbs the UV and retain the same even after the acetone and alcohol washing, which would interfere with the assay giving incorrect values of PHA concentration.

FTIR spectroscopy has been in use to characterize the intracellular PHA molecules which gives insight into the chemical structure without prior hydrolysis of the polymer molecule. This method facilitates rapid identification of the polymer and requires very less amount (0.5–1 mg) of sample. Rapid screening of scl and Mcl-PHAs in lyophilized bacterial cells can be done by using FTIR. For scl-PHAs the band at 1,185 cm^{-1} occurs due to C–O stretching and the band at 1,282 cm^{-1} corresponding to the CH group (Liebergesell et al., 1992). Similarly, for Mcl-PHAs the characteristic peak of ester carbonyl band occurs at 1,742 cm^{-1} and the band at 1,165 cm^{-1} which occurs due to C–O stretching (Mc Cool and Cannon, 2001).

GC-MS analysis is done for quantitation of PHA, including the total amount, mass and concentration of monomers present in the polymer at low concentrations of sample (Braunegg et al., 1978). In this method, the lyophilized cells were subjected to direct acid or alkaline methanolysis, followed by gas chromatography (GC) of the methyl esters. This method is fast, susceptible, and reproducible.

NMR spectroscopy such as 13C, 1H and 2D INADEQUATE (Incredible Natural Abundance Double Quantum Transfer Experiments) are in use for characterizing and quantifying the pure PHAs. By the method, the exact location of the double bonds in the monomer and the exact configuration of the polymer can be determined.

6.5 PHYSICAL PROPERTIES OF PHA

The physical properties of the polymer are mainly dependent on the chemical formulation and structure of the polymer itself. PHB is an optically active, compact helical structure, which is most like that of polypropylene, which also has a compact configuration. Characteristics of biopolymers are almost like those of the conventional plastics like polypropylene (Rehm et al., 1998; Schlegel et al., 1970; Ostle and Holt, 1982.). The MW of the bacterial polymers is in the range of 2×105–3×106 Da with a polydispersity index (Mw/Mn) of ~ 2. The properties crystallinity values of poly(3HB) are in the range of 55–80% (Kranz et al., 1997; Degelau et al., 1995; Williamson and Wilkinson, 1958). Poly(3HB) crystals usually show a lamellar morphology and form spherulites when crystallized from the melt to the bulk material. Poly(3HB) is a relatively stiff, rigid material and has a tensile strength comparable to that of polypropylene. Some of the scl-PHAs may be too brittle and rigid with lack of mechanical properties required for biomedical and packaging applications. Mcl-PHAs may be elastomeric and have low mechanical strength. Copolymers PHBV and Mcl-PHAs are less stiff and brittle, but most of the other mechanical properties are the same.

6.6 APPLICATIONS OF PHA

PHAs have been drawing considerable attention as biocompatible plastics for a wide range of applications due to its biocompatibility, biodegradability, and uniform chirality. Several different companies are developing PHAs for use in the plastic industry. Currently, some US and Japanese companies are actively trying for commercialization of PHA biopolymers. PHAs can also be used to make medical-surgical gears, carpets, packaging, compostable bags, lids, or tubs and flushable.

In recent years, nanoparticles (NPs) have been gaining importance for diverse biomedical applications like antioxidant, cytotoxic, antibacterial properties, etc., in the daily human life. Silver is used in different biomedical fields for therapeutic uses. Silver helps in the use of different liquid and solid compounds. The antibacterial property of the silver nanoparticles (SNPs) was well reported by different researchers. Many workers reported on the stability of the colloidal SNPs. Owing to the bactericidal properties of silver ions, its nanodispersions can form a basis for the development of new classes of bactericides, pesticides, and different medicinal substances. Surfactants like PVP (polyvinylpyrrolidone) and Tween, CTAB (Cetyl trimethylammonium 46 bromide) are used to stabilize SNPs. The natural hydrophobic polymers such as PHA from bacteria are expected to be an effective alternative renewable source for such uses. PHAs can be applied for:

- Blow molded bottles, containers, creating agent on paper and electronic products especially from the composites of PHA;
- PHAs have been studied as drug carrier scaffolds for controlled drug delivery because the physical properties like strength, modulus, and elongation of the scaffolds are comparable to that of other drug delivery systems. They are also used in medicines, insecticides, and herbicides;
- Due to the slow hydrolytic degradation inside the human body, PHAs can be more advantageous for their use in reconstructive surgery;
- On the basis of uniform chirality possessed, PHAs can also be used as the starting material for some other chemicals;
- PHAs have a potential role as a pollution bioindicator in preliminary assessments of environmental health (Rehm et al., 1998);
- In bone and blood vessel replacements; and
- In biosensor applications.

KEYWORDS

- **biosensor**
- **cetyltrimethylammonium 46 bromide**
- **gas chromatography**
- **polyvinylpyrrolidone**
- **silver nanoparticles**
- **sodium dodecyl sulfate**

REFERENCES

Asrar, J., et al., (2002). Biosynthesis and properties of poly (3-hydroxybutyrate-co-3-hydroxyhexanoate) polymers. *Biomacromolecules, 3*, 1006–1012.

Barham, P. J., (1990). Physical properties of poly (hydroxybutyrate) and poly (hydroxybutyrate-co-hydroxyvalerate). In: Dawes, E A., (ed.), *Novel Biodeg. Microbial Polymers* (pp. 81–96). Kluwer, Dordrecht.

Berger, E., et al., (1989). PHB recovery by hypochlorite digestion of non-PHB biomass. *Biotechnol. Tech., 3*, 227–232.

Braunegg, G., et al., (1978). A rapid gas chromatographic method for the determination of poly-β-hydroxybutyric acid in microbial biomass. *European J. Appl. Microbiol. Biotechnol., 6*, 29–37.

Braunegg, G., et al., (1998). Polyhydroxyalkanoates, biopolyesters from renewable resources: Physiological and engineering aspects. *J. Biotechnol., 65*, 127–161.

Degelau, A., et al., (1995). Fluorometric measurement of poly-b-hydroxybutyrate in *Alcaligenes eutrophus* by flow cytometry and spectrofluorometry. *Appl. Microbiol. Biotechnol., 42*, 653–657.

Hahn, S. K., et al., (1993). The recovery of poly (3-hydroxybutyrate) by using dispersions of sodium hypochlorite solution and chloroform. *Biotechnol. Tech., 7*, 209–212.

Hahn, S. K., et al., (1994). Optimization of microbial poly (3-hydroxybutyrate) recovery using dispersions of sodium hypochlorite solution and chloroform. *Biotechnol. Bioeng., 44*, 256–261.

Kim, Y. B., & Lenz, R. W., (2001). Polyesters from microorganisms. *Advances in Biochem. Engg. Biotech, 71*, 51–79.

Kranz, R. G., et al., (1997). Positive selection systems for discovery of novel polyester biosynthesis genes based on fatty acid detoxification. *Appl. Environ. Microbiol., 63*, 3010–3013.

Liebergesell, M., et al., (1992). Isolation and identification of granule-associated proteins relevant for poly (3-hydroxyalkanoic acid) biosynthesis in *Chromatium vinosum* D. *FEMS Microbiol. Lett., 78*, 227–232.

Mc Cool, G. J., & Cannon, M. C., (2001). PhaC and Pha R are required for polyhydroxyalkanoic acid synthase activity in *Bacillus megaterium*. *J. Bacteriol., 183*, 4235–4243.

Ostle, A. G., & Holt, J. G., (1982). Nile Blue A as a fluorescent stain for poly-b-hydroxybutyrate. *Appl. Environ. Microbiol., 44*, 238–241.

Pieper-Furst, U., et al., (1995). Identification of the region of a 14-kilodalton protein of *Rhodococcus ruber* that is responsible for the binding of this phasin to polyhydroxyalkanoic acid granules. *J. Bacteriol., 177*, 2513–2523.

Potter, M., et al., (2003). The complex structure of polyhydroxybutyrate (PHB) granules: Four orthologous and paralogous phasins occur in *Ralstonia eutropha*. *Microbiol., 150*, 2301–2311.

Qi, Q., & Rehm, B. H. A., (2001). Polyhydroxybutyrate biosynthesis in *Caulobacter crescentus*: Molecular characterization of the polyhydroxybutyrate synthase. *Microbiol., 147*, 3353–3358.

Ramsay, J. A., et al., (1994). Extraction of poly-3-hydroxybutyrate using chlorinated solvents. *Biotechnol. Tech, 8*, 589–594.

Rehm, B. H. A., et al., (1998). A new metabolic link between fatty acid de novo synthesis and polyhydroxyalkanoic acid synthesis. *J. Biol. Chem., 273*, 24044–24051.

Salehizadeh, H., & Loosdrecht, M. C. M. V., (2004). Production of polyhydroxyalkanoates by mixed culture: Recent trends and biotechnological importance. *Biotechnol. Adv., 22*, 261–279.

Schlegel, H. G., et al., (1970). The isolation of mutants not accumulating poly-b-hydroxybutyric acid. *Arch Microbiol., 71*, 283–294.

Spiekermann, P., et al., (1999). A sensitive, viable-colony staining method using Nile red for direct screening of bacteria that accumulate polyhydroxyalkanoic acids and other lipid storage compounds. *Arch. Microbiol., 171*, 73–80.

Williamson, D. H., & Wilkinson, J. F., (1958). The isolation and estimation of poly-beta-hydroxy butyrate inclusions of *Bacillus* species. *J. Gen. Microbiol., 19*, 198–200.

CHAPTER 7

Molecular Identification of Bacterial Isolates

Bacterial identification with phenotypic characters is so accurate as compared to genotypic methods. To estimate the relationship between the phylogeny and genotype determination of bacteria, the ribotyping is used as the most common molecular biological approach by comparing the similarities in the rRNA gene sequences. It is also commonly used as a rapid tool for the identification of unculturable and unknown bacteria up to the species level.

Ribotyping with DNA restriction fragments containing the genes coding for the 16S and 23S rRNA is reliable. 16S-rRNA genes are mainly used for classification and identification of bacteria. These genes are universally present in all bacteria and have highly conserved and variable regions that will discriminate the genus or species level. Moreover, 16S genes are located on the mitochondrion, and as a result numerous copies are available. This is adequately distinctive from its eukaryotic, archaeal, mitochondrial, and chloroplastic homologs, so that each can be amplified separately. Since it is enough conserved so almost universal bacterial amplification can be performed using specific primers. The 16S-rRNA gene is ~1,550 bp and composed of conserved and variable regions having interspecific polymorphism to provide distinguishing and statistically valid measurements. It is used as the template for PCR to amplify a gene having ~500–1,500 bp. Complementary universal primers to the conserved regions are usually chosen as they can be amplified from different bacteria. These primers are used at the 540 bp region at the up or downstream regions (~1,550-bp region), and the variable sequence

Bacterial Biopolymers. B. K. Konwar, PhD (Author)

within is used for comparative taxonomy. The other advantage of this gene is that for describing a new species, it is mandatory to sequence and deposit in an available site, so anyone can have access to the sequences of all described species and use them to identify any bacteria of interest. By digesting the 16S-rDNA gene with a restriction enzyme, fragments of variable lengths are generated. Gel electrophoresis with the restriction digested samples help in the visualization of fragments. The DNA bands reveal a unique pattern for each species and can also be used to identify the origin. Carl Woese first proposed this work and established the three-domain classification of Archaea, Bacteria, and Eucarya based on such sequence information.

For 16S-rRNA gene sequence identification of organisms the concerned databases, appropriate names associated and an accurate sequence of the isolate can be identified. The relationship of conserved 16S-rRNA gene sequence in *Bacillus* spp. was first stated by Law and Slepecky (1969). It marks the evolutionary relatedness of organisms. The comparison of 16S-rRNA gene sequences allows differentiation among organisms at the genus level of bacteria, classification of strains at multiple levels including species and subspecies. Other genomic regions have also been used to examine the phylogenetic relationships among bacteria but due to different genome sizes caused by duplication, transfer, deletion, fusion, and splitting of genes may occur. PHAs are the polyesters of various hydroxyalkanoates accumulated by a large variety of bacteria when, bacterial growth is limited by depletion of nitrogen, such as *Azotobacter* spp., phosphorus (Kansiz et al., 2000), or oxygen and an excess amount of carbon source is still present. If the general physiological fitness of bacteria is not affected, they can store excess nutrients inside their cells and by polymerizing soluble intermediates into insoluble molecules and the cells do not undergo alterations of its osmotic state. Peters and Rehm (2005) have been reported the prevention of leakage of these compounds from the bacterial cells and to allow the availability of the nutrient stores at a low maintenance.

7.1 PHA PRODUCTION IN BACTERIA

Azotobacter eutrophus bacteria with the diameter of 02–0.5 m contains 8–13 PHA granules. The granules appear to be highly refractive inclusion under electron microscopic observation. Braunegg et al. (1978) reported PHB granules in bacterial cells which can be detected by staining with Sudan Black B. However, Ostle and Holt (1982) advocated the use of Nile Blue A, a water-soluble basic oxazine dye that has a greater affinity and higher specificity than Sudan Black B for PHB, and that gives a bright orange fluorescence at a wavelength of 460 nm as other inclusion bodies do not stain with Nile Blue A, emphasizing its usefulness. The presence of PHB granules in live cells of *Caryophanon latum* could be confirmed by staining with Nile red at the early stage of accumulation closer to the cytoplasmic membrane as revealed by confocal laser scanning fluorescence microscopy (CLSFM). The PHA-storing bacteria in aerobic dynamic feeding broth of propionate as carbon source, was ascertained by RT-PCR in micromanipulated cells and confirmed by FISH (Lemos et al., 2008).

7.2 PHA PRODUCTION IN SPECIFIC BACTERIA

PHAs are produced by the different bacterial cultures. *R. eutropha* is mostly used bacteria for the production of PHB and PHBV. The species can produce PHB with the use of simple carbon sources. The strain NCIMB 11599 could produce 121 $g.L^{-1}$ PHB with a controlled glucose concentration (glucose, propionic acid and either 4-hydroxybutyric acid or butyrolactone) and nitrogen limitation in a fed-batch culture as carbon sources in fed-batch cultures producing P(3HB-4HB-3HV) terpolymer (Madden et al., 2000). Volova et al. (2005) reported *R. eutropha* B8562 produced PHA up to 90% of the cell dry weight by using different carbon sources (CO_2, fructose, and glucose). Because of the ability to use methanol as a carbon source, methylobacterium species are of interest for PHB production. *M. rhodesianum, M. extorquens, M. organophilum, Mrhodenium, Mzatmanii, M. radiotolerans, Mycoplana rubra, Paracoccus denitrificans* (Foollner et al., 1995) and *Proteomonas extorquens* (Ueda et al., 1992) are some of the bacteria which can produce PHB in methanol as the carbon source. Bourque et al. (1992) reported *M. extorquens* accumulated 30% (CDW) of P(3HB) with a molecular mass of 250,000 Da with methanol concentration of 1.7 gL^{-1}, and addition of complex nitrogen source. *P. putida* can efficiently incorporate monomers in the range of C_8C_{10} during PHA synthesis. De Smet (1983) observed the presence of intracellular granules of poly(3-hydroxyoctanoate) in *P. oleovorans* strain ATCC29347 grown in medium containing 50% v/v octane. Haywood et al. (1989) examined various *Pseudomonas* spp. for growth and polyester accumulation with C_6–C_{10} straight-chain alkanes, alcohols, and alkanoic acid as the carbon source. The accumulation of PHAs containing Mcl-3HB (C_6–C_{12}), but not 3-HB, appears to be a characteristic of the fluorescent *Pseudomonas* spp. to accumulate PHAs (Huisman et al., 1989). *P. putida* and *P. aeruginosa* are able to convert acetyl-CoA to Mcl monomers for PHA synthesis (Shang et al., 2003). Smibert and Krieg (1981) reported that *Pseudomonas* spp. 14-3 strain isolated from Antarctica accumulated large quantities of PHB in octanoate medium. This isolate was characterized based on partial sequencing of 16S-ribosomal RNA gene. The strain showed increased tolerance to both thermal and oxidative stresses.

Other important strains recently studied include: *Bacillus* spp., *Alcaligenes* spp., *Rhodopseudomonas palustris, Escherichia coli, Burkholderia* spp., and *Halomonas boliviensis*. The species *Rhodopseudomonas palustris* could produce PHB phototrophically outside in a temperature controlled, under-water bioreactor. Several *Bacillus* spp. has been reported to accumulate 9–44% dry cell weight of PHB. By comparison, *Bacillus* mycoides RLJ B-017 contained 69.4 ± 0.4% dry cell weight PHB. This strain could be potent industrially. The tropical marine and mangrove microbial seven isolates from the mid-west coast of India possessed a high potential to accumulate important polymers such as PHA with the accumulation of more than 1 gm of PHA per liter of culture broth (Rawte et al., 2002). Lemos et al. (2008) reported that a PHA accumulating gram-positive bacterium *Bacillus* spp. INT005 isolated from a gas field soil was identified by comparing morphological, physiological properties and partial nucleotide sequence ~500 bp of its 16S-rRNA. Construction of recombinant *B. subtilis* was also done to produce PHA co polymer (Wang et al., 2006).

Rhizobium spp. isolated from leguminous plants and the standard strains accumulated 27–57% PHA of their cell biomass while using sucrose as the sole carbon source (Lakshman and Shamala, 2003). Van-Thuoc (2008) produced PHB by utilizing *H. boliviensis* culturing in cheap and readily available agro-residue as cheap carbon source to reduce the production cost. The use of recombinant *E. coli* as PHA producer has become popular as it is genetically well characterized. *E. coli* not being a natural PHA accumulator, PHA production must be metabolically engineered in *E. coli* and it does not have any depolymerase activity to degrade the accumulated

PHA. Although *R. eutropha* produces high levels of P(3HB), but they have certain limitations which impede genetic manipulation. The PHA biosynthetic gene expression in *R. eutropha* and P(3HB) synthesis in *E. coli* for P(3HB) opened the possibility for the production by recombinant organisms (Schubert et al., 1988). Genetically and metabolically engineered *E. coli* can synthesize a variety of polymers, such as P(3HB-3HV), P(3HB-4HB), P(4HB) and P(3HO-3HH). Expression of *P. aeruginosa* PHA synthases, PhaC1 and PhaC2 in *E. coli* fadB mutant resulted in Mcl-PHA accumulation when grown in the presence of C_8–C_{14} fatty acids. Recombinant *E. coli* DH5 (pQKZ103) has been shown to accumulate PHB up to 85% of the cell dry weight while culturing in minimal glucose medium. PHA synthase gene (PhaC1) from indigenous *Pseudomonas* spp. strain LDC-5 was PCR amplified and cloned in *E. coli* is a potential candidate for the large-scale production of polymer (Langenbach et al., 1997).

7.3 PHA PRODUCING GENES

The conserved sequences were identified by using the specific primer-pair I-179L/I-179R for the identification of PHA biosynthetic genes PhaC1 and C2 genes. Solaiman et al. (2002) reported the primer pair, I-179L and I-179R which are highly specific for the coding regions of *Pseudomonas* PhaC1 and PhaC2 genes. The PCR product of 540 bp length revealed the presence of the coding regions of PhaC1 and C2 genes flanked by the primer pair. Solaiman (2000b) reported the semi-nested PCR reaction should specifically yield the 540 bp sub-genomic fragments of PhaC1 and PhaC2 on *P. resinovorans* NRRLB-2649 and *P. corrugata* strains. The result suggested the successful amplification of the specific coding regions and proceeded to confirm the production of Mcl-PHA by the *P. aeruginosa* strains JQ796859 and JQ866912. However, the identification of these PHA synthetic genes will need further confirmation by sequence determination, followed by restriction digestion study and transfer to specific bacteria for expression study.

7.4 GENOTYPIC CHARACTERIZATION

Since 16S-rRNA gene sequences are highly conserved and specific in all organisms and provide the degree of similarity among the groups of the organisms, indicating evolutionary relatedness, which helps to place an organism in the proper group. In this investigation, the phylogenetic analysis, indicated that the sequence similarity of 16S-rDNA of *P. aeruginosa* JQ796859 strain within the species was 99%, other closely related species of the strain was found to be *P. aeruginosa* P21 (GenBank entry: HQ697283) having a homology of 99.98%. Further, a BLAST (NCBI) search showed 99% homology to other known *P. aeruginosa* 16S-rDNA gene sequences and accordingly it was named as *P. aeruginosa* JQ796859. The phylogenetic analysis indicated a comparative search for the sequence of the strain *P. aeruginosa* JQ866912 revealed 99.98% homology to *P. aeruginosa* MTH8 (GenBank entry: HQ202541). Further, NCBI data-based BLAST search showed 99% homology to other known *P. aeruginosa* 16S-rDNA gene and accordingly named *P. aeruginosa* JQ866912 (GeneBank entry: JQ866912).

KEYWORDS

- ***Azotobacter eutrophus***
- ***Bacillus* spp.**
- **confocal laser scanning fluorescence microscopy**
- **electrophoresis**
- **gene sequences**
- **phylogenetic analysis**

REFERENCES

Bourque, D., et al., (1992). Production of poly-β-hydroxybutyrate from methanol: Characterization of a new isolate of *Methylobacterium extorquens*. *Appl. Microbial. Technol., 37*, 7–12.

Braunegg, G., et al., (1978). A rapid gas chromatographic method for the determination of poly-β-hydroxybutyric acid in microbial biomass. *European J. Appl. Microbiol. Biotechnol., 6*, 29–37.

Colak, A., & Guner, S., (2004). Polyhydroxyalkanoate degrading hydrolase-like activities by *Pseudomonas* sp. isolated from soil. *Int. Biodeter. Biodeg.,53*, 103–109.

De Smet, M. J., et al., (1983). Characterization of intracellular inclusions formed by *Pseudomonas oleovorans* during growth on octane. *J. Bacteriol., 154*, 870–878.

Foollner, C. G., et al., (1995). Biosynthesis of poly-3-hydroxybutyric acid by the facultatively methanol-assimilating bacterium *Mycoplana rubra* B346 and recombinant strains. *J. Basic Microbiol., 35*, 179–188.

Foster, L. J. R., et al., (2001). Environmental concentrations of polyhydroxyalkanoates and their potential as bioindicators of pollution. *Biotechnol. Lett., 23*, 893–898.

Haywood, G. W., et al., (1989). The importance of PHB-synthase substrate specificity in polyhydroxy- alkanoates synthesis by *Alcaligenes eutrophus. FEMS Microbiol. Lett., 57*, 1–6.

Holmes, P. A., (1985). Applications of PHB- a microbially produced biodegradable thermoplastic. *Phys. Technol., 16*, 32–36.

Huisman, G. W., et al., (1989). Synthesis of poly-3-hydroxyalkanoates is a common feature of fluorescent *Pseudomonads. Appl. Environ. Microbiol., 55*, 1949–1954.

Jendrossek, D., & Handrick, R., (2002). Microbial degradation of polyhydroxyalkanoates. *Annu. Rev. Microbiol., 56*, 403–432.

Kansiz, M., et al., (2000). Quantitative determination of the biodegradable polymer poly (b-hydroxybutyrate) in a recombinant *Escherichia coli* strain by use of mid-infrared spectroscopy and multivariate statistics. *Appl. Environ. Microbiol., 66*, 3415–3420.

Lakshman, K., & Shamala, T. R., (2003). Enhanced biosynthesis of polyhydroxyalkanoates in a mutant strain of *Rhizobium meliloti. Biotechnol. Lett., 25*, 115–119.

Langenbach, S., et al., (1997). Functional expression of the PHA synthase gene phaC1 from *Pseudomonas aeruginosa* in *Escherichia coli* results in poly (3-hydroxyalkanoate) synthesis. *FEMS Microbiol. Lett., 150*, 303–309.

Law, J., & Slepecky, R. A., (1969). Assay of poly-hydroxybutyric acid. *J. Bacteriol., 82*, 52–55.

Lemos, P. C., et al., (2008). Microbial characterization of polyhydroxyalkanoates storing populations selected under different operating conditions using a cell-sorting RT-PCR approach. *Appl. Microbiol. Biotechnol., 78*, 351–360.

Madden, L. A., et al., (2000). Production and characterization of poly (3-hydroxybutyrate-co-3-hydroxyvalerate-co-4-hydroxybutyrate) synthesized by *Ralstonia eutropha* in fed-batch cultures. *Polymer, 41*, 3499–3505.

Ostle, A. G., & Holt, J. G., (1982). Nile blue A as a fluorescent stain for poly-b-hydroxybutyrate. *Appl. Environ. Microbiol., 44*, 238–241.

Rawte, T., et al., (2002). Incidence of marine and mangrove bacteria accumulating polyhydroxyalkanoates on the mid-west coast of India. *World J. Microbiol. Biotechnol., 18*, 655--659.

Schubert, P., et al., (1988). Cloning of the *Alcaligenes eutrophus* genes for synthesis of poly-beta-hydroxybutyric acid (PHB) and synthesis of PHB in *Escherichia coli. J. Bacteriol., 170*, 5837–5847.

Shang, L., et al., (2003). Poly (3-hydroxybutyrate) synthesis in fed-batch culture of *Ralstonia eutropha* with phosphate limitation under different glucose concentrations. *Biotechnol. Lett.*, 25, 1415–1419.

Smibert, R. M., & Krieg, N. R., (1981). General characterization. In: Gerhardt, P., et al., (eds.), *Manual of Methods for General Bacteriology* (pp. 409–443). American Soc. Microbiol., Washington, DC.

Solaiman, D. K. Y., (2006b). PCR cloning of *Pseudomonas resinovorans* polyhydroxyalkanoates biosynthesis genes and expression in *Escherichia coli. Biotechnol. Lett., 22*, 789–794.

Solaiman, D. K. Y., et al., (2002). Physiological characterization and genetic engineering of *Pseudomonas corrugata* for medium-chain-length polyhydroxyalkanoates synthesis from triacylglycerols. *Curr. Microbiol., 44*, 189–195.

Ueda, S., et al., (1992). Synthesis of poly (3-hydroxybutyrate-co-3-hydroxyvalerate) from methanol and n-amyl alcohol by the methylotrophic bacteria *Paracoccus denitrtificans* and *Methylobacterium extorquens*. *Appl. Environ. Microbiol., 58*, 3574–3579.

Volova, T. G., et al., (2005). Physiological and biochemical characteristics and capacity for polyhydroxyalkanoates synthesis in a glucose-utilizing strain of hydrogen-oxidizing bacteria, *Ralstonia eutropha* B8562. *Microbiol., 74*, 684–689.

Wang, Y., et al., (2006). Construction of recombinant *Bacillus subtilis* for production of polyhydroxy alkanoates. *Appl. Biochem. Biotechnol., 129–132*, 1015–1022.

CHAPTER 8

The Fermentation Process

The production of PHA from bacteria can be augmented with the culture of concerned bacteria species is the need of the hour. PHA-producing bacteria are classified in to two groups based on culture conditions required for the efficient PHA synthesis. One group requires the limited supply of essential elements like N, P, Mg, O, or S, whereas the other group requires no nutrient limitation and could synthesize and accumulate polymer during growth. Therefore, these characteristics can be considered in developing culture methods for the efficient production of PHAs. Either fed-batch or continuous cultivation techniques can be used for producing PHA with high productivity. The nutritionally enriched growth phase yields sufficient biomass with product formation in nitrogen-depleted medium. Single fed-batch fermentation that is nitrogen limited lead to low amounts of PHA, as there is not enough accumulation of biomass (Slater et al., 1988).

The use of activated sludge can contribute to decrease the cost of PHAs and therefore increase their commercial potential. A stable methane-utilizing mixed bacterial culture used in the large-scale production of PHB using cheap substrates like methane from natural or renewable sources in an open system. It has become possible to increase PHB content to 87% by using nitrogen limited fed-batch culture, which is higher than that of obtainable 50% under nitrogen-available conditions (Choi and Lee, 1997). Further experimentations need to be carried out to see whether continuous cultivation can truly give higher productivity than fed-batch cultures without any process problems, such as, culture instability and contamination. A stress-induced system was developed by Kang et al. (2008) with PHB biosynthetic pathway, in which fermentation was caused by recombinant *E. coli* DH5 (pQKZ103) harboring the system capable to accumulate PHB up to 85.8% of cell dry weight in minimal glucose medium without the addition of inducer. PHA production based on mixed microbial culture was tried to decrease the cost of production, having no requirement of sterilization as bacteria could adapt well to the was developed with the use of excess carbon substrate for a short period, followed by prolonged starvation (Majone et al., 1996) for the production of PHB. The intracellular PHA content reached 65.4% cell dry weight under ADF conditions with the addition of carbon by sequential pulses (Serafim et al., 2004).

In the bacterial cell, carbon substrates are metabolized by 3 different pathways: Pathway I generate PHB homopolymer from acetyl-CoA processed from sugars and has been found in bacteria like *R. eutropha*. Pathways II and III generate Mcl-(R)-3HA monomers mostly from fatty acid oxidation and fatty acid biosynthesis intermediates (Green et al., 2002), respectively which are mostly found in various fluorescent *Pseudomonas spp*. Xi et al. (2000) reported the production of PHA up to 77% of cell dry weight by *Pseudomonas stutzeri* strain 1317 in media containing fatty acids, alcohols, diols, glucose, and gluconate. The synthesis of random co-PHAs by *P. aeruginosa* ATCC 27853 with the use of Tween 20 as the mixed carbon source was obtained. Tween 20 and its three major fatty acids support both cell growth and PHA production and its emulsifying and solubilizing properties seems to facilitate its use as a carbon source by the bacterial cells, leading to the production of the highest polymer yield. Moreover, it could provide fatty acids substrates at a lower cost than that of purified fatty acids (Impallomeni et al., 2000). A native strain of *B. cereus* CFR06 was able to produce amylase and polyhydroxyalkanoate (PHA) 48% CDW in the starch containing medium (Halami, 2008).

The different kinds of agricultural wastes like starch (Poomipuk et al., 2014) beet and cane molasses (Page, 1992), corn syrup, wheat bran (Obruca et al., 2015), malt waste (Wang et al., 2007), and dairy wastes like cheese whey (Bosco and Chiampo, 2010) with or without nitrogen supplement were used as raw materials for the production of PHA. Efforts have been made to grow bacteria on different renewable vegetable oils and various waste products (Verlinden et al., 2011). The inexpensive carbon sources to produce PHAs could lead to significant economical advantage. Hahn et al. (1999) reported *Haloferax mediterranei*, an archaeon capable of producing PHA 55.6 and 38.7 weight percent in the repeated fed-batch fermentation with the use of low-cost substrate like extruded rice bran (ERB) and extruded corn starch (ECS), respectively. *P. putida* can produce medium chain length PHA (28 gL^{-1}) from an inexpensive and renewable carbon source, corn oil hydrolysate by a fed-batch culture (Shang et al., 2008). Liu et al. (1998) reported production of 39.5 $gL^{-1}h^{-1}$ PHB with the use of molasses as the source of carbon in the case of recombinant *E. coli* strain. Plant oils or their derived fatty acids are good carbon source to produce PHA as they are inexpensive renewable carbon sources. Large-scale production of P(3-HB-co-3-HHx) in lauric acid containing media was obtained with the use of *Aeromonas hydrophila* with a final PHA content of 50%.

Novel processes like use of organic wastes, waste mater (Pratt et al., 2019; Yu, 2001), industrial wastes (Kumar et al., 2004) and municipal wastes (Lee and Yu, 1997; Lee et al., 2000b) have been investigated to produce PHAs. The different wastes containing volatile fatty acids (VFAs) like palm oil mill effluent (Loo et al., 2005), banana pseudostem (Kalia et al., 2000), damaged food grains, pea shells, apple pomace have been used as carbon sources. Bacteria like *A. eutrophus*, *B. megaterium, P. oleovorans, Azotobacter* sp., *Beijerinckia* spp., *Rhizobium* spp., *Nocardia* spp., etc., utilize food wastes as substrate for PHA production. The production from organic wastes could be beneficial to the environment and could promote sustainable development. Some PHA-producing microbes like *R. eutropha* cannot directly utilize organic wastes as they are usually in the complex form. So, the first step to overcome this problem is hydrolysis and acidogenesis of the wastes producing volatile acetic, propionic, and butyric acids, which could be used by *R. eutropha* for synthesis of P(HB-co-HV).

CO_2 in the atmosphere is the ultimate feedstock for PHA production. Some wild-type cyanobacteria can accumulate small amounts of P(3-HB) (approx. 6%) in the cells from CO_2. *R. eutropha* can assimilate CO_2 and produce P(3-HB) in the absence of light energy, but with oxidization of hydrogen. CO-resistant bacteria *R. eutropha* strain B5786 can synthesize PHAs up to 75% under the autotrophic conditions (Ishizaki et al., 2001). *Cupriavidus necator* H16 can metabolize a mixture of H and CO_2 to form PHAs (Verlinden et al., 2007).

The use of cheap carbon source methanol as the substrate is significant for the production of PHB. It can also be considered as a renewable substrate since it could be derived from woody materials or from natural gas obtained after anaerobic digestion of organic substances. The strain GW2 of *Methylobacterium* spp. isolated from groundwater was found to be able to produce the homopolymer P(3HB) in methanol, ethanol, and succinate. The strain showed the best PHB production, 40% w/w dry biomass in methanol containing medium (Bourque et al., 1992).

The recombinant *E. coli* bacteria harboring PHA biosynthetic gene of *A. latus* is able to produce high concentration of P(3-HB-co-3-HV) with 3-HV fractions in fermentation. *A. latus* when transformed with its own cloned PhaC gene it exhibited increased PHB synthesis as well as increase in PHB content. The recombinant *A. latus* synthesized the maximum PHB concentration from 3.1–3.7 gL^{-1} and its content from 50.2–65% of cell dry weight, respectively, as compared to the untransformed *A. latus* (Choi and Lee, 1997).

8.1 THE RECOVERY PROCESSES

The high cost due to maintenance of pure bacterial cultures and use of organic substrates, the recovery and production of PHA polymer become expensive. Several methods have been suggested for the economic recovery of purified PHA. The bacterial cells with PHAs following fermentation are separated by centrifugation, washed, dried, and disrupted to recover the polymer. Most of the methods to recover purified intracellular PHA involve the use of chloroform, methylene chloride, propylene carbonate or dichloroethane. The solvent extraction method is good for the medical application as it gives a higher percentage of purified PHAs. The use of cheaper and less toxic solvents such as hexane, acetone, and dimethyl carbonate for the PHA recovery process is desirable. The polymer solution having above 5% (w/v) P(3HB) is viscous; as such the removal of cell debris becomes difficult. Moreover, the process needs large quantities of toxic-volatile chemicals, increasing the cost of production and causing environmental hazard (Koller et al., 2013). The digestion using sodium hypochlorite is an alternative to the unfavorable extraction with organic solvents. Even though this method is effective in the digestion, it causes PHA degradation. Surfactant pretreatment followed by hypochlorite digestion results in pure isolation of P(3HB) with less degradation and improved molecular weight (MW). The use of enzymatic digestion to produce Biopol with expensive chemicals and complex processes does not seem to be economically feasible (Ramsay et al., 1994). Fidler and Dennis (1992) reported a system for PHB recovery from *E. coli* cells by expressing T7 bacteriophage lysozyme gene. In this system, the lysozyme penetrated and disrupted the bacterial cells resulting PHB granules to be released. Choi and Lee (1999) reported a simple alkaline digestion method for the recovery of P(3HB) from the recombinant *E. coli* cell culture. The recombinant cells having a P(3HB) content of 77% on treatment with 0.2 M NaOH for 1 h, P(3HB) of 98.5% purity. A simple two-step process was developed by Hampson and Ashby (1999) to extract and purify Mcl-PHAs from *P. resinovorans* cells. In this method, the lipid impurities are removed by supercritical fluid extraction of the lyophilized cells followed by chloroform extraction using CO_2. By this method maximum of 42.4% Mcl-PHA was obtained referring to time saving by this process, use of much less organic solvent, and production of purer cl-PHA biopolymer.

In most of the organisms so far investigated, the PHB is synthesized from acetyl-coenzyme A (acetyl-CoA) by a sequence of three reactions catalyzed by three biosynthetic enzymes: In (i) 3-ketothiolase (Pha A) combines two molecules of acetyl-CoA to form acetoacetyl-CoA, (ii) the acetoacetyl-CoA reductase (Pha B) allows the reduction of acetoacetyl-CoA by NADH to 3-hydroxybutyryl-Co A. Finally (iii) PHB synthase (Pha C) polymerizes 3-hydroxybutyryl-CoA to PHB and coenzyme A being liberated (Figure 5.4). During the normal bacterial growth, the 3-ketothiolase will be carried over by the free coenzyme A coming out of the Krebs cycle. The re-entry of acetyl-CoA into Krebs cycle is restricted, the surplus acetyl-Co A is channeled in to PHB biosynthase (Uchino et al., 2007). In *Rhodospirillum* rubrum, two stereospecific enoyl-Co A hydratases are also involved (Moskowitz and Merrick, 1969); these enzymes catalyze the conversion of L-(+)-3-hydroxybutyryl-Co A via crotonyl-Co A to D-(–)-3-hydroxybutyryl-Co A, which is polymerized to yield PHB.

KEYWORDS

- ***azotobacter* sp.**
- **coenzyme**
- **extruded corn starch**
- **extruded rice bran**
- **Krebs cycle**
- **polyhydroxyalkanoates**
- ***Rhizobium* spp.**

REFERENCES

Bosco, F., & Chiampo, F., (2013). Production of polyhydroxyalkanoates (PHAs) using milk whey and dairy wastewater activated sludge: Production of bioplastics using dairy residues. *J. Biosci. Bioeng., 109*, 418–421.

Bourque, D., et al., (1992). Production of poly-β-hydroxybutyrate from methanol: Characterization of a new isolate of *Methylobacterium extorquens*. *Appl. Microbial. Technol., 37*, 7–12.

Choi, J., & Lee, S. Y., (1997). Process analysis and economic evaluation for poly (3-hydroxybutyrate) production by fermentation. *Bioprocess Eng., 17*, 335–342.

Choi, J., & Lee, S. Y., (1999). Efficient and economical recovery of poly (3-hydroxybutyrate) from recombinant *Escherichia coli* by simple digestion with chemicals. *Biotechnol. Bioeng., 62*, 546–553.

Du, G., Si, Y., & Yu, J., (2001). Inhibitory effect of medium-chain-length fatty acids on synthesis of polyhydroxyalkanoates from volatile fatty acids by *Ralstonia eutropha*. *Biotechnol. Lett., 23*(19), 1613–1617.

Fidler, S., & Dennis, D., (1992). Polyhydroxyalkanoate production in recombinant *Escherichia coli*, *FEMS Microbiol. Lett., 103*, 231–235.

Green, P. R., et al., (2002). Formation of short chain length/medium chain length polyhydroxyalkanoate copolymers by fatty acid β-oxidation inhibited *Ralstonia eutropha*. *Biomacromol., 3*(1), 208–213.

Halami, P. M., (2008). Production of polyhydroxyalkanoate from starch by the native isolate *Bacillus cereus* CFR06. *World J. Microbiol. Biotechnol., 24*, 805–812.

Hampson, J. W., & Ashby, R. D., (1999). Extraction of lipid-grown bacterial cells by supercritical fluid and organic solvent to obtain pure medium chain-length polyhydroxyalkanoates. *JAOCS, 76*, 1371–1374.

Huang, T. Y., et al., (2006). Production of polyhydroxyalkanoates from inexpensive extruded rice bran and starch by *Haloferax mediterranei*. *J. Ind. Microbiol. Biotechnol., 33*, 701–706.

Impallomeni, G., et al., (2000). Tween 20 and its major free fatty acids as carbon substrates for the production of polyhydroxyalkanoates in *Pseudomonas aeruginosa* ATCC 27853. *J. Polym. Environ., 8*, 97–102.

Ishizaki, A., et al., (2001). Microbial production of poly-D-3-hydroxybutyrate from CO_2. *Appl. Microbiol. Biotechnol., 57*, 6–12.

Kalia, V. C., et al., (2000). Anaerobic digestion of banana stem waste. *Bioresour. Technol., 73*, 191–193.

Kang, Z., et al., (2008). Construction of a stress-induced system in *Escherichia coli* for efficient polyhydroxyalkanoates production. *Appl. Microbiol. Biotechnol., 79*, 203–208.

Koller, M., et al., (2013). Strategies for recovery and purification of poly [(R)-3-hydroxyalkanoates] (PHA) biopolyesters from surrounding biomass. *Engg. Life Sci., 13*(6), 549–562.

Kumar, M. S., et al., (2004). Production of biodegradable plastics from activated sludge generated from a food processing industrial wastewater treatment plant. *Bioresour. Technol., 95*, 327–330.

Lee, S., & Yu, J., (1997). Production of biodegradable thermoplastics from municipal sludge by a two-stage bioprocess. *Resour. Cons. and Recycl., 19*, 151–164.

Liu, F., et al., (1998). Production of poly-b-hydroxybutyrate on molasses by recombinant *Escherichia coli*. *Biotechnol. Lett., 20*, 345–348.

Loo, C. Y., et al., (2005). Biosynthesis and characterization of poly (3-hydroxybutyrate-co-3-hydroxyhexanoate) from palm oil products in a *Wautersia eutropha* mutant. *Biotechnol. Lett., 27*, 1405–1410.

Moskowitz, G J., & Merrick, J. M., (1969). Metabolism of poly-β-hydroxybutyrate. II. Enzymic synthesis of D-(-)-β-hydroxybutyryl coenzyme A by an enoyl hydrase from *Rhodospirillum rubrum. Biochem., 8*, 2748–2755.

Obruca, S., et al., (2015). Use of lignocellulosic materials for PHA production. *Chem. Biochem. Engg. Quart., 29*(2), 135–144.

Page, W. J., (1992). Production of polyhydroxyalkanoates by *Azotobacter vinelandii* UWD in beet molasses culture. *FEMS Microbiol Lett., 103*, 149–157.

Patnaik, P. R., (2005). Perspectives in the modeling and optimization of PHB production by pure and mixed cultures. *Crit. Rev. Biotechnol., 25*, 153–171.

Poomipuk, N., et al., (2014). Poly-β-hydroxyalkanoates production from cassava starch hydrolysate by *Cupriavidus* sp. KKU38. *Int. J. Biol. Macromol., 65*, 51–64, 2014.

Pratt, S., et al., (2019). Polyhydroxyalkanoate (PHA) bioplastics from organic waste. In: *Biorefinery* (pp. 615–638).

Ramsay, J. A., et al., (1994). Extraction of poly-3-hydroxybutyrate using chlorinated solvents. *Biotechnol Tech., 8*, 589–594.

Serafim, L. S., et al., (2004). Optimization of polyhydroxybutyrate production by mixed cultures submitted to aerobic dynamic feeding conditions. *Biotechnol. Bioeng., 87*, 145–160.

Shang, L., et al., (2008). Mass production of medium-chain-length poly (3-hydroxyalkanoates) from hydrolyzed corn oil by fed-batch culture of *Pseudomonas putida*. *World J. Microbiol. Biotechnol., 24*, 2783–2787.

Slater, S. C., et al., (1988). Cloning and expression in *Escherichia coli* of the *Alcaligenes eutrophus* H16 poly-beta-hydroxybutyrate biosynthetic pathway. *J. Bacteriol., 170*, 4431–4436.

Steinbuchel, A., & Lutke-Eversloh, T., (2003). Metabolic engineering and pathway construction for biotechnological production of relevant polyhydroxyalkanoates in microorganisms. *Biochem. Engg. J., 16*, 81–96.

Uchino, K., et al., (2007). Isolated poly (3-hydroxybutyrate) (PHB) granules are complex bacterial organelles catalyzing formation of PHB from acetyl coenzyme A (CoA) and degradation of PHB to acetyl-CoA. *J. Bacteriol., 189*(22), 8250–8256.

Verlinden, R. A. J., et al., (2007). Bacterial synthesis of biodegradable polyhydroxyalkanoates. *J. Appl. Microbiol., 102*, 1437–1449.

Verlinden, R. A. J., et al., (2011). Production of polyhydroxyalkanoates from waste frying oil by *Cupriavidus necator*. *AMB Express, 1*, 1–8.

Wang, Y. J., et al., (2007). Synthesis of PHAs from waster under various C: N ratios. *Biores. Technol., 98*(8), 1690–1693.

Xi, J., et al., (2000). Hyperproduction of polyesters consisting of medium-chain-length hydroxyalkanoate monomers by strain *Pseudomonas stutzeri* 1317. *Antonie Van Leeuwenhoek, 78*, 43–49.

Yu, J., (2001). Production of PHA from starch wastewater via organic acids. *J. Biotechnol., 86*, 105–112.

CHAPTER 9

Structure and Properties of PHAs

9.1 THE STRUCTURE

Although the microbial origin, structure, and physicochemical properties of PHAs is quite variable (Figure 2), the repeat scl-PHAs units are hydroxy fatty acids and the pendant group (R) varies in between methyl (C_1) to tridecyl (C_{13}). Fatty acids with the OH group at position 4, 5, or 6 and R containing substitutes or unsaturation are also known. Within bacterial metabolism, carbon substrates are converted to hydroxyacyl-CoA thioesters. In all characterized PHAs so far, the OH substituted carbon is of the stereochemical configuration. Few bacteria can produce scl-PHAs containing monomers other than 3HB when grown on simple sugar (Moejko-Ciesielska and Kiewisz, 2016). According to Sato et al. (2005) the most common PHB homopolymers, synthesized by bacteria, always contain less than 1 mol of 3-HV monomers. Some *Pseudomonas* spp. are reported to accumulate PHA copolymers containing Mcl monomers. The composition of PHA produced is usually related to the substrate used for growth mostly 2n carbon shorter than the substrate used (Ojumu et al., 2004). The use of mixed substrates like glucose and valerate causes the change in the conversion of copolymers of PHB to PHBV or PHB4B (Pederson et al., 2006). If the use of substrates is alternated overtime, there is a possibility to obtain PHA block copolymers synthesized in bacterial culture. Labuzek and Radecka (2001) observed that the gram-positive *B. cereus* UW85 strain could produce tercopolymer by using 3-HB, 3-HV, and 3 hydroxyhexanoate units of caprolactone and glucose as carbon source. The bacterial species *R. eutropha* synthesizes copolymer of 3-PHB and

Bacterial Biopolymers. B. K. Konwar, PhD (Author)

3-mercapto propionate like poly(3-HB-co-3MP) containing sulfur in the backbone if 3-mercaptopropionic or 3,3-thiodipropionic acid is provided in addition to fructose or gluconic acid in the culture medium under nitrogen-limited conditions (Vijay and Tarika, 2018). In contrast to the normal production of copolymer while using the mixed substrates in the culture of *Pseudomonas nitroreducens* isolated from oil-contaminated soil demonstrated the unusual ability to synthesize PHB homopolymer from Mcl-fatty acids including hexanoate and octanoate (Dalal et al., 2010).

The surface of PHA granule is coated with a layer of phospholipid and proteins. Phasins, a class of predominant protein of the PHA granule influence the number and size of it (Potter and Steinbuchel, 2005). *Pseudomonas* putida strain CA-3 could accumulate polyphosphate (poly P) and it appeared that poly P was not the rate-limiting step for Mcl-PHA accumulation in *Pseudomonas* strains (Tobin et al., 2007).

9.2 PHYSICAL PROPERTIES OF PHAS

Among all the biopolymers, PHB are the most widely studied polymer and therefore, the properties of PHAs have been elucidated (Van de Velde and Kiekens, 2002). The average molecular weight (MW) of scl-PHAs, like PHB produced from wild-type bacteria is mostly in the range of 1×10^4–4×10^6 Da with a polydispersity value of ~2. But according to Valappil et al. (2007), the MW of Mcl-PHAs lies between the range of 6×10^4 and 4.12×10^5. The polydispersity index of scl-PHAs and has to be ~ 1.75, whereas Mcl-PHA copolymers in the range of 1.6–4.4 (Rai et al., 2011).

The family of PHA could be hard crystalline to elastic depending on the compositional mass of monomers extending its application. Scl-PHAs like P(3HB) homopolymer are highly crystalline, stiff, and brittle material (Loo and Sudesh, 2007). When spun into fibers, it behaves as a hard-elastic material (Singh, 2015). The glass transition (Tg) and melting temperature (Tm) is 180°C and 4°C, respectively. The polymers become fluid and can be molded above their melting temperature. Mechanical properties, Young's modulus (3.5 GPa) and tensile strength (40 MPa) are close to that of polypropylene, although the elongation to break is about 5%, which is significantly lower than polypropylene 400%. According to Ashby et al. (1998) the film of Mcl-PHA synthesized by *Pseudomonas resinovorans* from coconut oil, tallow, and soybean oil can be improved when subjected to 50 kGy irradiation. It results in the formation of crosslinks of olefinic groups present in the polymer side chains and improves the tensile strength (104% and 63%), elongation (49% and 13%) and Young's modulus (30% and 76%) of the polymer film.

The Mcl-PHAs have their mechanical properties similar to PHB, but they are less stiff and brittle. By changing copolymer compositions both physical and thermal properties can be regulated. Due to the low melting temperature (40–60C) and low glass transition temperature (–50°C to –25°C, Mcl-PHAs are semi-crystalline elastomers. Sanchez et al. (2003) reported Mcl-PHAs having low crystallinity due to the presence of large and irregular pendant side groups. The copolymer of 3HV becomes tougher and more flexible with the increase of 3HV units. However, the melting temperature decreases and elongation to break becomes more with increasing 3HV fraction. The melting behavior and crystallization of PHAs show multiple melting peak behavior and melting-recrystallization-remelting (Gunaratne and Shanks, 2005). Carrasco et al. (2006) reported that the thermal biodegradation of PHB (Biopol) starts at 246.3°C, while the value for PHBV (Biopol) is 260.4°C as the side chain increases in the later which in turn increases the thermal stability. The presence of valerate in the polymer chain increases the thermal stability of the polymer. The introduction of co-monomers other than 3HV into PHB gives improved mechanical properties to copolymers. The physical properties of some polymers are compared with polypropylene and polystyrene in Table 9.1.

TABLE 9.1 Physical Properties of Some Biopolymers

Polymer	Melting Point (°C)	Glass-Transition Temperature (°C)	Crystalline (%)	Young's Modulus (GPa)	Tensile Strength (MPa)	Elongation to Break (%)
P(3HB)	179	4	80	3.5	40	6
P(4HB)	53	–	–	149	104	1,000
P(3HB-co-3HV)	–	–	–	–	–	–
3 mol% 3HV	170	–	–	2.9	38	–
9 mol% 3HV	162	–	–	1.9	37	–
14 mol% 3HV	150	–	–	1.5	35	–
20 mol% 3HV	145	–	–	1.2	32	–

TABLE 1 *(Continued)*

Polymer	Melting Point (°C)	Glass-Transition Temperature (°C)	Crystalline (%)	Young's Modulus (GPa)	Tensile Strength (MPa)	Elongation to Break (%)
25 mol% 3HV	137	–1	40	0.7	20	–
P(3HB-co-HB)	–	–	–	–	–	–
3 mol% 4HB	166	–	–	–	28	45
10 mol% 4HB	159	–	–	–	24	242
16 mol% 4HB	150	–7	45	–	26	444
64 mol% 4HB	50	–	–	30	17	591
90 mol% 4HB	50	–	–	100	65	1,080
P(3HB-10% 3HHx)	127	–1	34	–	21	400
Polypropylene (PP)	176	–10	50–70	1.7	34.5	400
Polystyrene	110	21	–	3.1	50	–

Source: Data adapted from: Poirier et al. (1995); Lee (1995); Tsuge (2002); Khanna and Srivastava (2005); Verlinden et al. (2011).

9.3 PHYSICAL ANALYSIS OF PHA PROPERTIES

9.3.1 MOLECULAR WEIGHT (MW)

The molecular weight (MW) of P-3HV-5-HDE isolated from the bacterial strain *P. aeruginosa* JQ796859 was 5.6×10^3–5.9×10^3 Da. A narrow polydispersity index value (I = Mw/Mn) 1.059 suggests the presence of uniform chain length which is essential for food and pharmaceutical industries as there is need for further processing towards polymer chain uniformity. The average MW of P-3HV-5-DE is considerably smaller than other reported Mcl-PHA polymers. The molar mass is almost one order smaller than Mcl-PHA polymers isolated from *P. putida* (MW 33,000 Da and MW 56,000 Da) whereas the polydispersity index was slightly higher (1.7).

In the case of *B. circulans* strain MTCC8167, the molecular mass of PHBV was revealed to be 5.1×10^4 Da by GPC and 5.2×10^5 Da of PHB whereas PHBV has a very PDI of 1.21 (Oliveira et al., 2007). The PDI is comparable to PHB 4 isolated from the recombinant *R. eutropha*. In the case of copolymer 3-hydroxyoctanoate and 3-hydroxydecanoate from *P. putida* strain KT2047A was reported to be 1.24 by Ma et al. (2009). The low PDI of PHBV suggests its use for commercial purpose.

TP-3HV-5HDE-3HODE of *P. aeruginosa* JQ866912 possesses MW 3.8×10^4 Da, whereas the average MW of the polymer is 4.1×10^4 Da with PDI 1.08. *P. putida* KT2442 produced Mcl-PHA containing hydroxy-dodecanoate fraction with the average MW of 8×10^4 Da with slightly higher PDI (1.25) than the polymer of *P. aeruginosa* JQ866912; Allen et al. (2010) *P. oleovorans* (ATCC 29347) grown in saponified jatropha seed oil produced copolymer with MW 1.79×10^5 and PDI 1.3 almost similar to *P. aeruginosa* JQ866912 and suggested that the low PDI refers to uniform chain length.

9.3.2 CRYSTALLINITY STUDY OF PHA POLYMER FROM BACTERIAL STRAINS

Polymers of *P. aeruginosa* JQ796859 and JQ866912, *B. circulans* MTCC8167 were crystalline in nature. The X-ray diffraction (XRD) pattern supported that the polymer obtained from the solvent casting of the first strain was crystalline possessing about 32.17% crystallinity, P-3HB-co-3HV polymer of *B. circulans* MTCC8167 strain showed crystal size of 2.260 Å and the polymer showed 65% of crystallinity, but the crystallinity of P3HB of *R. eutrophus* in 4-hydroxybutyric acid was reported to be 59% by Abe et al. (1994). The XRD to determine the strain out of the polymer chains by using Voigt-function model was described by de-Keijser et al. (1982) and the value between crystals was found to be 36.69%. The polymer from *P. aeruginosa* JQ866912 strain also showed crystallinity 46%.

9.3.3 THERMOGRAVIMETRIC (TGA) ANALYSIS

The TGA graph revealed that the standard polymer from the market possessed the lowest thermal degradation temperature in comparison to all the polymers. From the TGA thermogram the polymer from *P. aeruginosa*

JQ796859 showed that the first step thermal degradation was 225°C and about 90% of the degradation of the neat PHA occurred at 385°C indicating high degree of temperature resistivity of the polymer. The polymer of *P. aeruginosa* JQ866912 and *B. circulans* MTCC8167 was thermally stable up to 220–229°C. The first 1.5% of weight loss step in TG analysis curves was at around 80–120°C corresponding to the moisture loss and volatilization of the solvent. The second step of 80% of weight loss in the TGA curves at 350–360°C was caused by degradation of polymers. Only 10% residue of all polymers was finally left on heating to 600°C. The Mcl-PHA is more thermostable whereas PHB and P(3HB12 mol% 3HHx) degraded at 226 and 239°C, respectively as reported by Liu and Chen (2007). High thermal degradability and melting temperatures of the isolated PHA from these strains revealed the applicability of Mcl-PHAs in different industries. The Mcl-PHAs consisting of 30 mol% 3HV, the degradation temperature was between 296°C and 307°C (Li et al. (2007)). PHAs from fermented molasses and single volatile fatty acids (VFAs) in open mixed culture produced were 277.2°C and 294.9°C degradation temperature, respectively.

9.3.4 DIFFERENTIAL SCANNING CALORIMETRY (DSC) ANALYSIS

DSC analysis of polymer from the bacterial strain *P. aeruginosa* JQ796859 revealed the Tm value to be about 139°C. A second melting peak at around 227°C indicated the presence of a second monomer. DSC of a copolymer on incorporation of second monomer possessed two melting temperatures. The polymer was melt-stable on heating above the Tm to 190C, quenched to–30°C, and reheated to 300°C; the peak at the Tm was obtained again on second heating, suggesting their stability. This Tm value can be compared with the Tm value of the polymer isolated from other *Pseudomonas* sp. reported using glucose as carbon source. The Tm of P3HB homopolymer of *Azotobacter chroococcum* 23 was 180°C and Tg of 3.9°C being similar to the glass transition temperature; Tg of PHA from *P. aeruginosa* JQ796859 (4°C) as reported by Li et al. (2007). The polymer was also showing some segmental melting of polymer chain at around 34–55°C. In the case of *B. circulans* MTCC8167 polymer, the DSC graph revealed the Tg to be 14.79°C. The melting temperature of the polymer was about 116°C and the second at around 234°C. The Tm data was like Tm data (116.4°C) of the P(3HB) with 20 mol.% 3HV produced by *A. chroococcum*. They also stated that the Tg of the polymers and melting range decreases with the increased inside chain length. At around 28.7°C, the polymer was also showing some segmental melting of the polymer chain. The Tg of the polymer from *P. aeruginosa* JQ866912 strain was found to be 1.03°C. Here the first melting point (Tm) starts at around 150°C and second melting peak at around 226°C. This thermal behavior of the polymer from *P. aeruginosa* JQ866912 was similar to the Tm (155°C) and Tg (1.2°C) of the polymer produced with fermented molasses with 44 mol% non-3HB by an open mixed culture. No ashes have been found, in the thermal degradation of all these polymers and the polymers are of copolymer in nature.

9.3.5 PHOTOLUMINESCENCE (PL) STUDY

The emission was recorded at two different wavelengths for all three polymers. All polymers showed luminescence property; the emission maxima (PL max) of polymers are found within 340–390 nm. This value was similar to the fluorescence spectra of glycogen biopolymer from bovine liver reported by Bozani et al. (2011).

9.3.6 SURFACE MORPHOLOGY STUDY OF PHA BIOPOLYMER

Scanning electron micrograph (SEM) of the polymer sample of *P. aeruginosa* JQ796859 and JQ866912 strains possessed a coralloid surface which was decorated with many stars shaped protrusions. In the case of *B. circulans* MTCC8167 polymer, the film failed to show the star shaped protrusions and the polymer film possessed two non-uniform layers. But in both the cases, there was the absence of pore in the polymer film surface. Thire et al. (2006) have reported that the SEM micrograph of pure PHB film has fractured surface with relatively smooth topography, indicating a brittle fracture holding several cracks and voids.

9.3.7 CELL SIZE DETERMINATION BY SEM ANALYSIS

The determination of the shape and size of PHA producing bacterial strains was done by SEM and the photographs thereof are presented in Figure 9.1.

The SEM micrograph has revealed all bacterial strains to be rod shaped and each bacterial cell appeared 1–2 μm in size.

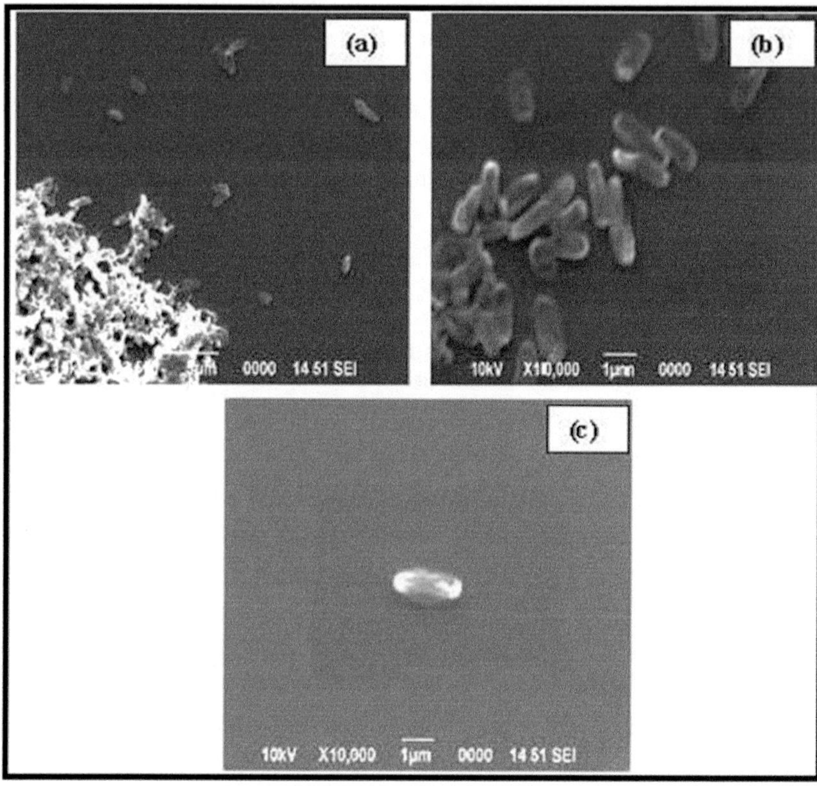

FIGURE 9.1 SEM micrograph of bacterial strains. (a) *P. aeruginosa* JQ796859; (b) *B. circulans* MTCC8167; and (c) *P. aeruginosa* JQ866912. A bacterial cell appears 1–2 μm in size.

9.4 BIOPOLYMER PRODUCING BACTERIA

A total 13 potential bacterial polyhydroxyalkanoate (PHA) producing isolates belonging to *Pseudomonas* aeruginosa and *Bacillus circulan* were obtained. Dalal et al. (2010) reported the isolation of different bacterial strains from soil samples contaminated with oily sludge procured from the different oil refineries for biopolymer PHA production. The environment of petroleum hydrocarbon containing with excess carbon coupled with nitrogen stress, can be a potential source for most the bacteria for the production of PHA. Various workers (He et al., 1998; Kaplan and Kitts, 2004; Haba et al., 2007) reported that *Pseudomonas* species is the most exploited bacteria producing PHAs from oil-contaminated soils. Reddy and Thirumala (2012) also reported potent PHA-producing bacterial strain from crude oil contaminated ecosystem.

9.4.1 MORPHOLOGICAL CHARACTERISTICS OF BACTERIA

Morphological studies of bacteria in terms of colony size, pigmentation, form, margin, and elevation displayed wide variations. The size of colonies varies from large, moderate, small, and pinhead. In most of the cases, colonies are white, at times yellowish and green also seen. Difference in morphology exists among bacterial communities at the crude oil contaminated surface and sub-surface soil which could be influenced by the complex environmental factors.

9.4.2 BIOPOLYMER PRODUCING BACTERIA

In the investigation out of 13 bacterial isolates BPC1 (*P. aeruginosa* JQ796859), BPC2 (*P. aeruginosa* JQ866912) and BP2 (*B. circulans* MTCC8167) showed positive-growth. Soil bacteria can be screened in PHA detection medium (Rehman et al., 2007). These isolates were promising for PHA production. Emission of orange color under the fluorescence microscope confirmed the three isolates to be PHA producer. The isolates also showed bluish black color colonies on staining with the Sudan Black B. Ostle and Holt (1982) reported emission of orange fluorescence by PHB granules on staining with 1% Nile blue A; but not by glycogen and polyphosphate. In comparison to Sudan black B, Nile blue A stain is more specific to PHB. William and Christopher (1996) reported the presence of inclusion bodies of PHB inside *Azotobacter vinelendi* UWD cells producing fluorescence emission on staining with Nile blue A. Staining with Sudan black B and Nile blue A can be used in screening of *Enterobacter* spp. SEL2 and *Enterobacteriaceae bacterium* PFW1 as reported by Naheed et al. (2012). PHB in a variety of bacteria cell can be used for the quantitative assay by these stains.

9.4.3 BIOCHEMICAL CHARACTERIZATION OF PHA PRODUCING BACTERIA

The gram-negative *P. aeruginosa* JQ796859 exhibited catalase, citrate, nitrate reduction activity whereas lack of urease, triple sugar iron, MR, and VP activity. The microbe failed to show starch utilization, H_2S (hydrogen sulfide) and indole production. The strain possessed the ability to hydrolyze gelatin and casein and also the ability to peptonize the milk protein. The isolate could produce acid from lactose, dextrose, sucrose, and xylose but not from mannitol. The isolate JQ866912 possessed catalase, citrate, and triple sugar iron activity but none for MR and VP. The isolate showed the ability to reduce nitrate to nitrite and could utilize glucose; but failed to show ability in the utilization of starch and urease, production of H_2S and indole. The microbe could peptonize milk protein and hydrolyze gelatin and casein. Except lactose, it could show positive result for acid production from mannitol, dextrose, sucrose, and xylose.

The gram-positive spore-forming *B. circulans* MTCC8167 showed catalase, urease, triple sugar iron activity but none in the case of MR-VP. The isolate could produce acid from lactose, mannitol, dextrose, sucrose, and xylose, but failed to reduce nitrate to nitrite as well as uptake of glucose. The bacterium hydrolyzed gelatin and casein; and possessed the ability to utilize triple sugar iron and peptonized milk. The microbe did not show any activity for the citrate, H_2S and indole production, but could hydrolyze starch.

9.4.4 GROWTH OF BACTERIAL STRAINS AT DIFFERENT PH LEVELS

Each microbial species has its own definite pH range and growth optimum. Some microorganisms can survive within a wide range of pH, while others can only endure small variations. Microorganisms can be affected due to plasma membrane disruption or inhibition of active enzymes and membrane transport proteins by excessive variations in H ion concentration. In the investigation, at high acidic (pH 2) condition, growth of the bacterial isolates was poor with (0.25–0.34 gL^{-1}) dry biomass production. In between 6 and 8, the bacterial isolate *B. circulans* MTCC8167 possessed better growth with comparatively higher biomass yield to the other two isolates of *P. aeruginosa*, JQ796859 and JQ866912. At pH 10, JQ796859 possessed better growth though other bacteria showed lack of growth. Garland (1977) reported that in bacteria, intracellular pH is maintained in near neutral range, but when the cell is subjected to an acidic environment, the intracellular pH can decrease considerably. Russell and Dombrowski (1980) reported some enzymes are markedly sensitive to pH and their growth inhibition occurs at low pH that could be caused by the direct effect of H ion on cellular components.

9.4.5 GROWTH OF BACTERIAL STRAINS AT DIFFERENT TEMPERATURES

The PHA producing 3 bacterial isolates on culture in glucose containing medium; *P. aeruginosa* JQ796859 and JQ866912, *B. circulans* MTCC8167 exhibited better growth rate and dry biomass yield of 0.86–0.94 $g.l^{-1}$ at 30°C, but at 37°C the isolate JQ866912 exhibited the highest biomass yield of 4.3 gL^{-1} as compared to the other two. However, at a high temperature of 40°C, all bacterial strains failed to survive. The death is due to the enzyme-catalyzed reactions being sensitive to temperature. High temperatures can damage microorganisms by denaturing their enzymes, transport carriers, proteins, and thus cell membrane disrupted.

9.4.6 PHA PRODUCTION BY THE BACTERIAL STRAINS AT DIFFERENT GROWTH PHASES

The rate of production and accumulation of PHA by the bacterial strains in PHA detection medium having 1% glucose was maximum at 72 h when leading to stationary phase. There was no accumulation of PHA by all three bacterial strains in the first 12 h. The strain *P. aeruginosa* JQ866912 followed by *B. circulans* MTCC8167 exhibited the highest PHA accumulation after 48 to 72 h of culture. However, all three bacterial cultures exhibited a reduction in the accumulated PHA after 96 h of culture. According to Zinn et al. (2001) PHA accumulating microbes cultured in medium with carbon excess and nitrogen limitation, the cells must grow maximum up to the available nutrient in the medium. After that when stationary phase has reached, the microbes stopped their growth and accumulate PHA inside their cells as an energy reserve material for the future.

9.4.7 PHA PRODUCTION BY THE BACTERIAL STRAINS IN DIFFERENT CARBON SUBSTRATES

The bacterial strains were tested for PHA production with the different carbon substrates like glucose, glycerol, Colocasia starch, propionic acid, soy oil, waste mobile, sugarcane molasses and glycerol biproduct kitchen chimney dump lard (KCDL). PHA accumulation by the bacterial strain in glucose containing medium was the highest in *P. aeruginosa* JQ866912. Similar results were obtained in the glycerol containing medium. But, in the case of Colocasia starch, *P. aeruginosa* strains JQ796859 and JQ866912 possessed no growth; in the combination of 0.1% propionic acid and 1% glucose as the carbon source, the highest PHA accumulation by *B. circulans* MTCC8167; in sugarcane molasses and soy oil by JQ866912 whereas in waste mobile by MTCC8167; in the KCDL JQ866912 showed the highest PHA accumulation. Higher cell density was obtained by JQ866912 strain when glycerol byproduct was used. In the case of Colocasia starch both strains of *P. aeruginosa* and in waste mobile *B. circulans* MTCC8167 showed no growth. The poly(3-hydroxybutyrate-co-3-hydroxyhexanoate) (PHBHHx) was produced by genetically engineered *P. putida* GPp104 in the use of gluconate and glucose but not in fatty acids. Solaiman et al. (2002) obtained higher cellular yields than those obtained with *P. corrugata* in E medium supplemented with glucose (1.52 g CDW/l) or oleic acid (1.62 g CDW/l. In glucose or oleic acid containing medium, the two strains produced low yields of 31% and 61% CWD as compared to *P. corrugate* which can be compared with the higher accumulation of Mcl-PHA in *P. oleovorans* NRRL B-778 grown on glucose, octanoic acid or oleic acid (Ashby et al., 2002). The specificity of enzymes involved in PHA synthesis tend to determine the composition of polyester; the metabolically linked genetics of fatty acid metabolism to PHA synthesis and on the composition of the substrate. *P. oleovorans* NRRL B-14682 and *P. corrugata* synthesized Mcl-PHA up to 13–27% and 42% with cellular population up to 1.3 ± 0.1 gL^{-1} and 1.7–2.1 gL^{-1}, respectively in soy-based biodiesel production containing glycerol, fatty acid soaps or residual fatty acid methyl esters (FAME) as a fermentation feedstock. The culture of *Staphylococcus epidermidis* in sesame oil caused PHA accumulation of 2.5 gL^{-1} and cell density of 2.68% as reported by Wong et al. (2000). An efficient production of co-polymers was observed by Obruca et al. (2010) in waste oil as compared to pure oil and glucose. *Cupriavidus necator* produced 1.2 gL^{-1} PHB homopolymer in the heated frying oil or waste media similar to glucose. Unsaturated fatty acids, residual carbohydrates, proteins, and fats from foods, available nitrogen compounds, peroxides, heat-degraded waste or frying oil could also be used to metabolize and contribute to increased PHB production. The glycerol byproduct from KCDL could cause PHA accumulation better by the bacterial strains in comparison to the other carbon sources.

9.5 PHYSICOCHEMICAL CHARACTERIZATION OF PHA

9.5.1 LIGHT ABSORPTION AND FTIR ANALYSIS OF PHA

Three bacterial strains *P. aeruginosa* JQ796859, JQ866912 and *B. circulans* MTCC8167 showed maximum absorption (max) in the UV region (260–290 nm). Moreover, the peaks were not sharp enough. The PHA isolated from all the three strains contained the main characteristic ester carbonyl (C = O) group which was confirmed by the FTIR spectral pattern (Figure 9.2).

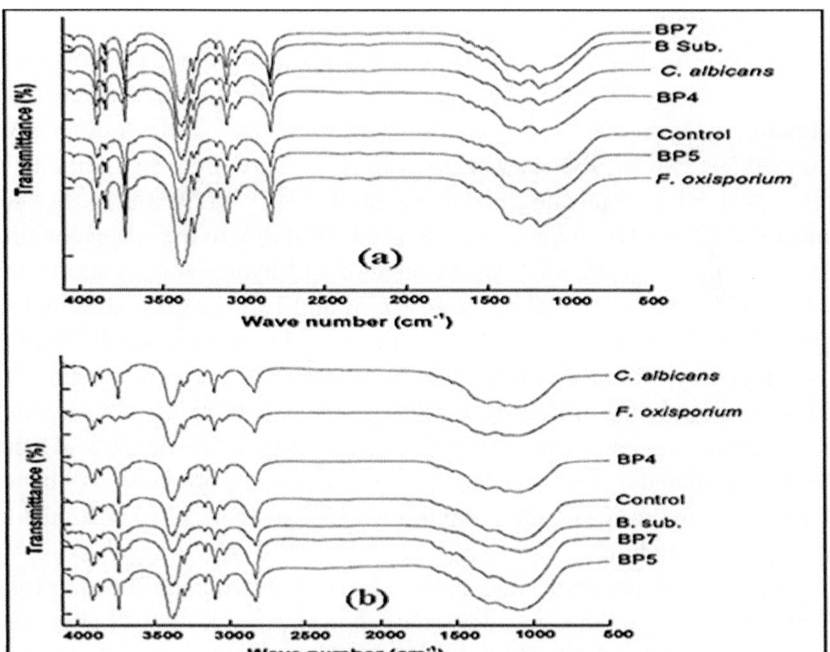

FIGURE 9.2 FTRI spectra of the biodegradation of PHA from *B. circulans* MTCC8167 isolate (a) after 15 days; and (b) after 36 days of treatment of microbes.

The peak between 3,405 cm^{-1} and 3,369 cm^{-1} might be attributed in the polymer due to O–H bonding. The peaks between 2,923 cm^{-1} and 2,922 cm^{-1} might be ascribed to the bands of C–H stretching in all three polymers isolated. The peaks ranged from 1,065–1,071 cm^{-1} might be due to COC stretching vibrations in the polymer material in all three strains. In the case of JQ796859 and JQ866912 polymer, the peaks ranged from 1,670–1,634 cm^{-1} might be ascribed to the C = C stretching. The polymers from these two strains JQ796859 and JQ866912, there was a presence of alkene CH stretching in the range of 3,085–3,055 cm^{-1}. The presence of a peak for the ester carbonyl bond indicated the accumulation of PHA. Moreover, these transmission bands at the above ranges were identical to those reported earlier for PHA (Misra et al., 2000).

9.5.2 GC-MS ANALYSIS OF POLYHYDROXYALKANOATES (PHA) FROM THE BACTERIAL STRAINS

PHA polymer can be subjected to qualitative and quantitative analysis by GC-MS for the actual monomeric characterization. In the bacterial strain *P. aeruginosa* JQ796859, the structural confirmation of the PHA monomers was done using GC-MS which revealed the production of a copolymer containing hydroxyvalerate and hydroxydecanoate in the polymer of JQ796859. Two methyl ester peaks with RT = 11.45 min, an additionally large peak RT = 19.65 and some minor ones at RT = 12.59 and 14.81 were observed. These two major peaks showed similarity to the corresponding standard mass spectra of hydroxyvalerate and hydroxydecanoate in the MS (NIST) library. In Figures 10.9, 10.11, and 10.13, methyl ester in GC spectra of the saturated 3-hydroxy valeric acid was dominated at m/z = 74, caused by McLafferty Rearrangement (1959) of the methyl ester. The expulsion of methanol from m/z = 270 probably caused m/z = 143. Similarly, the methyl ester mass spectra of the unsaturated decanoic acid were being dominated by m/z = 74 and 143. Therefore, the production of copolymer poly-hydroxyvalerate-co-hydroxydecanoate could be confirmed in this growth condition.

The isolated polymer's GC pattern possessed two methyl ester peaks with me RT = 22.67 min, an additionally large peak at RT = 28.1 and some minor ones at RT = 3.49, 24.95, and 30.91. The mass spectra of the methyl ester of the saturated 3-hydroxy butyric acid were dominant at m/z = 103 with a cleavage between C_3 and C_4. Further, important ions at m/z = 61 was probably of the methyl ester, but m/z = 203, would not be indicative of a special functionality in the molecule. Similarly, the methyl ester in the saturated 3-hydroxyvelerate were being dominated by m/z = 270. Further, important ions at m/z = 74, caused by McLafferty rearrangement of the methyl ester. m/z = 143 was probably caused by expulsion of methanol from m/z = 270. The other ions were not indicative. Therefore, it was concluded that a copolymer of poly-3-hydroxybutyrate-co-3-hydroxyvalerate (PHB-HV) was produced under this growth condition.

The GC pattern of the total ion chromatogram (TIC) of polymer from *P. aeruginosa* JQ866912 revealed three methyl ester peaks at RT = 11.56 min, 18.46 and additionally large peak at RT = 19.66 and minor peaks at RT = 12.59 and 14.61. The methyl ester of saturated 3-hydroxy valeric acid in mass spectra was being dominated by m/z = 103, effected by a cleavage between C_3 and C_4 and M/Z = 143 probably caused by the expulsion of methanol. Similarly, the decanoic acid methyl esters mass spectra pattern showed a peak at m/z = 90.1 which originated from the carbonyl end. The cleavage between C_2 and C_3 carbon with McLafferty rearrangement caused this. The methyl esters of octadecanoic acid methyl esters fragmentation pattern showed the peak at m/z = 74.1. The peak originated from the carbonyl end due to the cleavage between C_3 and C_4 carbon atoms following McLafferty rearrangement. The cleavage of bonds between C_3 and C_4 resulted in –OH end at the peak m/z 103.1, the alkyl end of this cleavage resulted in the peak at m/z = 71.1. It could be concluded that the poly-hydroxyvalerate-co-hydroxydecanoate-co-octadecanoate copolymer was produced by *P. aeruginosa* JQ866912 strain.

9.5.3 NMR ANALYSIS OF POLYHYDROXYALKANOATES (PHA) FROM THE BACTERIAL STRAINS

The H^1 and C^{13} NMR analyzes of the polymer in CDC13 solution was done to reveal the structure of the extracted polymer from *P. aeruginosa* JQ796859 strain. The chemical shift (Gottlieb et al., 1997) values in H^1 NMR ranging from 0.91–0.98 ppm attributed the protons of methyl terminal groups, while values at 1.22–2.08 ppm corresponded to all internal CH_2 groups in the polymer. The peak at 4.30–4.71 ppm was due to the presence of OCH (ester) proton and at 5.56–5.68 ppm clearly indicated CH = CH moiety.

The C^{13} NMR peaks appearing at 10–30 ppm highlighted the presence of methyl and methylene carbons. C^{13} signal at 128 and 129 ppm indicated CH = CH carbons and at 172–173 ppm exhibited peaks for two types of C = O carbons of the copolymer. From NMR data, it could be established that the polymer isolated from JQ796859 is poly(3-hydroxyvalerate) co-(5-hydroxydecanoate) co-(P-3HV-5-HDE. The integration height in H^1 NMR spectrum could determine the composition in P-3HV-5-HDE and it was found that the molar % composition of 3HV and 5-HDE was 45% and 55%, respectively.

In *B. circulans* MTCC8167 H^1 NMR and C^{13} NMR supported the structure proposed of the extracted polymer. In the H^1 NMR chemical shift values () ranging from 0.83 ppm represented the methyl terminal group protons, whereas values at 1.22–1.95 ppm represented internal CH_2. The presence of OCH proton at the peak 3.9–4.33 ppm and values at 5.3–5.5 ppm clearly indicated the CH = CH moiety in the polymer chain. Likewise, C^{13} NMR peak appearing at 1,030 ppm represented methyl and methylene carbons, at 129 and 130 ppm indicated CH = CH and that at 173.34–173.61 ppm presented representative peaks for two types of C = O carbons of the copolymer. Referring to the above, a medium chain length copolymer structure was proposed as P-3HB-3HV.

In the 1H NMR, chemical shift values ranging from 0.91–0.99 ppm attributed the protons of methyl terminal groups, while values at 1.22–1.93 ppm corresponded to all internal CH_2 groups in the polymer. The peak at 4.29–4.72 ppm was due to the presence of ester (OCH) proton and at 5.25–5.68 ppm clearly indicated CH = CH moiety in the polymer chain. Likewise, C^{13} NMR peaks appearing at 10–30 ppm represented methyl and methylene carbons. The C^{13} signal at 124.32–139.31 ppm indicated CH = CH carbons and that at 173.38–173.66 ppm peaks for two types of C = O of the copolymer composition. Presence of CH = CH moiety in the polymer chain indicates unsaturation in one monomer of the polymer. The spectrum closely resembles the spectra of the PHA isolated from *P. putida* cultivated on glucose and PHA isolated from *P. aeruginosa* MTCC 7925 with different carbon sources (Huijberts et al., 1992). From the NMR information, it could be proposed that the polymer isolated from *P. aeruginosa* JQ866912 contain 3-hydroxyvalerate (3HV), 5-hydroxydecanoate (5HDE) and 3-hydroxyoctadecanoate (3HODE).

9.6 BIODEGRADATION OF PHA

The stereochemistry of PHA is convenient for its biodegradability; CO_2 and organic compound are released in to the ecosystem which due to recycling that acts as buffer to climate change. Biopolymers are degraded both under aerobic and anaerobic conditions. They can also be subjected to thermal degradation and enzymatic hydrolysis. For the biodegradation of PHAs factors such as stereo-regularity, molecular mass, monomers, and crystallinity are accountable as influenced by the chemical structure like the presence of functional groups in the chain. Nishida and Tokiwa (1993) observed that the development of crystallinity could evidently depress the microbial degradability of PHB. A variety of factors like surface area, microbial activity in the disposal environment, pH, temperature, moisture, and presence of nutrient materials determine the rate of degradation of biopolymers. The surface morphology tends to encourage degradation through the contact between water, enzyme or bacteria and polymer chains. When exposed to hydrolysis, the degradation starts on the surface and at physical lesions on the polymer and proceeds to the inner part of the material.

The biodegradation of PHA should clearly distinguish between intra- and extra-cellular degradation. A non-accumulating microorganism causes the extracellular degradation that secretes extracellular PHA depolymerases. The source of extracellular polymer is the PHA released by accumulating cells after their death and cell lysis. PHA depolymerases secreted by few microorganisms hydrolyze the ester bonds into water-soluble monomers and oligomers to be transported into a microbial cell for metabolization to CO_2 and water (Wang et al., 2004).

Gram positive and negative bacteria, some methanogenic bacterial co-cultures have been reported to degrade PHA polymer by various researchers (Budwill et al., 1992). Marine Streptomyces sp. SNG9 utilizes PHB and its copolymer P(3HB-co-HV) as the sole source of carbon as reported by Mabrouk and Sabry (2001) and degrades in four days. Microorganisms like *P. lemoignei, P. pseudomallei, Acidovorax facilis, A. delafieldii, Comamonas testosteroni, Variovorax paradoxus, Zoogloea ramigera, Bacillus* sp. and *Streptomycetes* can degrade P(3HB) extracellularly (Jendrossek and Handrick, 2002). Three *Pseudomonas* spp. viz. fluorescens, aeruginosa, and putida isolated from fuel oil contaminated soil could degrade PHAs causing morphological changes as reported using SEM by Colak and Guner (2004). Phithakrotchanakoon et al. (2009) reported that a thermophilic *Streptomyces* sp. BCC23167 isolated from a landfill site is capable of degrading aliphatic polyesters like PHA copolymers, poly(-caprolactone) and polybutylene succinate at 50C and neutral pH. Another soil bacterium *Actinomadura* sp. AF-555 has the potential to degrade P-3HB-co-HV when exposed to soil (Shah et al., 2010).

Conversely intracellular degradation is the active degradation of an endogenous storage reservoir by the accumulating bacterium itself. Enzymes catalyzing the intracellular degradation of PHA are intracellular depolymerases. In the case of intracellular degradation, PHAs are degraded by some PHA producing bacteria intracellularly; in fact, the polymer is broken down to acetyl-CoA which under the aerobic condition enters the citric acid cycle and oxidized to CO_2. The PHA depolymerase enzyme phaZ is involved in the degradation of the PHAs (Knoll et al., 200). PHAs can be degraded in an anaerobic environment such as sewage sludge. It was reported that P(3HB) and the copolymer P(3HB-co-3HV) were fermented (up to 83–96%) to methane and CO_2 within 16 days by using anaerobically digested domestic sewage-sludge consortium (Mas-Castella et al., 1995). Lee (1995) reported that P(HB-HV) could not be degraded under the normal storage conditions as it is insoluble in water abut stable only in air. PHAs can be composted over a range of temperatures with the maximum at around 60°C. UV radiation also affects the degradation of bacterial biopolyester PHBHHx with significant molecular loss taking place with the UV radiation time (Shangguan et al., 2006).

Fungi play a considerable role in degrading bacterial biopolyesters due to its higher population density in soil. Fungi collected from environmental samples could degrade PHB in culture media as well. *Penicillium*, *Cephalosporium*, *Paecilomyces*, *Trichoderma*, *Ascomycetes*, *Basidiomycetes*, *Deuteromycetes*, *Mastigiomycetes*, *Myxomycetes*, and *Aspergillus fumigates* were identified to be PHA degraders, *Candida albicans* and *Fusarium oxysporum* were reported for the biodegradation of Mcl-PHA (Kim and Rhee, 2003).

9.6.1 BIODEGRADATION PROCESS OF BIOPOLYMERS

Biodegradation of the prepared biopolymer films of *B. circulans* MTCC8167 strain was carried out by treating with soil bacteria like *P.*

aeruginosa strains BP4, BP5, BP7, *B. subtilis* (R38-I) and fungal strains *Candida albicans* and *Fusarium oxysporum*. Biodegradation study of *P. aeruginosa* JQ796859 and JQ866912 biopolymer (PHA) was carried out by treating with *Alcaligens faecalis* (MTCC8164), *B. circulans* (MTCC8167), *P. aeruginosa* (MTCC7815) and *Mycobacterium* spp. (G-35I). At first, the biopolymer film was incubated in carbon-free mineral salt medium and then the bacterial and fungal strains were inoculated separately in each of the incubated Erlenmeyer flasks. The biopolymer films exhibited slow degradation after 15–36 days of inoculation. The increase in the microbial (bacterial/fungal) biomass on the surface of the PHA film and in the medium indicated that they utilized PHA as the source of carbon and energy.

The FT-IR spectrum of the PHA film of the *B. circulans* MTCC8167 strain inoculated with the bacterial and fungal strains showed shifting and decrease in intensity of the peak in comparison to the control, polymer without microbial treatment. Optical microscopic observation revealed many changes in the surface morphology of the PHA film of *B. circulans* MTCC8167 at the end of the incubation. From the data, it could be concluded that the biopolymer from bacterial strain *B. circulans* MTCC8167 is biodegradable.

The PHA films of *P. aeruginosa* JQ796859 and JQ866912 inoculated with each of the bacterial strains in FT-IR spectra also showed shifting and decrease in intensity of the peak in comparison to the control. This revealed the degradation pattern of the polymer.

The SEM of biodegradation of the polymer films of *B. circulans* MTCC8167, before and after 36 days is presented. The morphology of polymer surface support the degradation of the polymer when exposed to microbial action. Similarly, SEM micrograph of the PHA films of the bacterial strains *P. aeruginosa* JQ796859 and JQ866912 at the end of the incubation period of 36 days are presented. The micrographs also showed many changes in the surface morphology which depicts the biodegradation of the polymer samples.

Biodegradation of the biopolymer films was done by treating with soil bacterial and fungal strains. FTIR spectra of the PHA film of *P. aeruginosa* JQ796859, JQ866912 and *B. circulans* MTCC8167 strains was analyzed following the microbial exposure that indicated the formation of new peaks, breakdown of some bonds and increase in the band intensity which might be due to the production of monomers. The decrease in the intensity of bands at 1,127–1,140 cm^{-1} assigned to the stretching vibration of COC group in MTCC8167 indicated oxidative breakage of the polymer chain. Similarly, the decrease in the intensity of the bands around 11,851,211 cm^{-1} assigned to the stretching vibration of the COC group in the strains JQ796859 and JQ866912 polymer, indicated degradation of the polymer chain. Changes also occurred in the region of 1,675–1,735 cm^{-1} of the MTCC8167 polymer associated with C = O stretching of ester carbonyl bond with shifting and decrease in the peak intensity revealing breakdown of the films after treatment with the different microbes (Savenkova et al., 2000). Similarly, the PHA films of JQ796859 and JQ866912 strains inoculated with each bacterial strain possessed a shifting in the peak from 1,730–1,735 cm^{-1} associated with C = O stretching ester carbonyl bond referring to a decrease in the peak intensity which might correspond to the breakdown of ester bond in the polymer films following the inoculation. This was probably due to the externally secreted enzymes by microbes in the culture medium.

The FTIR spectra of The PHA film of all the bacterial strains had disappearance of a peak found in the control at 2,922 cm^{-1} and 2,919–2,923 cm^{-1} respectively, indicating the breakdown of CH bonds by exo-cleavage activity of PHA depolymerase. The breakdown proved decrease in amorphous nature due to decrease in the intensity of the peak between 764 cm^{-1} and 775 cm^{-1} and the peak between 722 cm^{-1} and 724 cm^{-1}. The shape and intensity of IR bands of the copolymer are affecting the crystallinity degree, and decrease in intensity of some bands, but the increase responds to higher crystallinity (Bhatt et al., 2008). The change is same in the case of synthetic polymers polyethylene and polyurethane before and after the microbial treatment as reported by FTIR spectroscopy. Bordoloi and Konwar (2009) also reported degradation of petroleum derived different complex hydrocarbons by the biosurfactant produced by the bacterial strains of *P. aeruginosa* MTCC7815, MTCC7812, L43-I, *B. subtilis* R38-I by reducing surface tension. We have used the four above *Pseudomonas* and *Bacillus* sp. and surfactant produced by them might be the reason for the degradation. Both bacterial and fungal species can potentially be used for the degradation of PHA. Minor bond degradation of some polymers enhances the degradation process via changing the crystallinity of polypropylene which otherwise are resistant to microbial decomposition, though there could be the appearance of cracks caused by the contraction of surface layers in semi-crystallization.’

SEM of the original films and after 36 days of biodegradation for the polymer of JQ796859, JQ866912 and MTCC8167 could change the surface morphology on being acted by microbial action. Films showed surface erosion in different areas, grooves, cavities, and result in pit formation on being subjected to microbial degradation compared to the original film. The cavities and grooves on the surface of PHBV films demonstrated synergistic degradation effect of a microbial population of fungi, bacteria, and actinobacteria colonization. The degradation profile of polyurethane/epoxy blends by different *P. aeruginosa* strains was reported by Dutta et al. (2010) with SEM micrograph. The action of exo- and/or endo-enzymes of the microbes themselves cause the degradation of polymer film. The accessible amorphous phase and chain ends at the crystal edges are initially attacked by enzymes in the case of PCL films. Accordingly, it was concluded that the biopolymer from MTCC8167 and *P. aeruginosa* JQ796859 and JQ866912 bacterial strains is biodegradable though it needs modified procedures and other parameters. Optical microscopic observation showed that the PHA film of bacterial strain *B. circulans* MTCC8167 at the end of incubation showed many changes in surface morphology, such as erosion and extensive roughening with pit formation as compared to the untreated one.

The breakdown of complex PHA compound into simpler compounds because of the action of microorganisms without the release of any hazards to the environment is desirable. Different types of biodegradations occur in the environment such as Hydro-biodegradation, Photo-biodegradation, and Oxo-biodegradation.

The intracellular degradation occurs in the endogenous storage reservoir by the accumulating bacterium itself and intracellular PHA depolymerases catalyze the degradation process. On the contrary, extracellular degradation is the utilization of an exogenous polymer released by accumulating cells after death. The different microbes present in the environment secreting extracellular PHA depolymerase and PHA hydrolase enzymes (Jendrossek, 2002) to hydrolyze solid PHA into water-soluble oligomers, and then utilize them as nutrients within the cells. After degradation of PHA, the polymer is converted into carbon dioxide and water under aerobic conditions and into methane under anaerobic conditions. In biological systems PHAs can be degraded using microbial depolymerases as well as by non-enzymatic and enzymatic hydrolysis in animal tissue (Lim et al., 2005). Factors like molecular mass, monomeric composition, chemical structure, crystallinity, etc., determine the biodegradation of PHAs. The rate of biopolymer biodegradation depends on various factors, like surface area, microbial activity of the environment, pH, temperature, moisture, and pressure of other nutrient materials. Ultraviolet ray can also accelerate the degradation of PHAs (Sridewi et al., 2006).

KEYWORDS

- **molecular mass**
- **molecular weight**
- **polyhydroxyalkanoates**
- **thermogravimetric**
- **volatile fatty acids**
- **X-ray diffraction**

REFERENCES

Abe, H., et al., (1994). Physical properties and enzymatic degradability of poly (3-hydroxybutyrate) stereoisomers with different stereoregularities. *Macromol., 27*, 6018–6025.

Allen, A. D., et al., (2010). Biosynthesis and characterization of copolymer poly (3HB-co-3HV) from saponified *Jatropha curcas* oil by *Pseudomonas oleovorans*. *J. Ind. Microbiol. Biotechnol., 37*, 849–856.

Ashby, R. D., et al., (1998). Improved film properties of radiation treated medium-chain-length poly (hydroxyalkanoates). *Biotechnol. Lett., 20*, 1047–1052.

Ashby, R. D., et al., (2002). The synthesis of short- and medium- chain-length poly (hydroxyalkanoate) mixture from glucose- or alkanoic acid-grown *Pseudomonas oleovorans*. *J. Ind. Microbiol. Biotech, 28*, 147–153.

Bengtsson, S., et al., (2010). Molecular weight and thermal properties of polyhydroxyalkanoates produced from fermented sugar molasses by open mixed cultures. *J. Biotechnol., 147*, 172–179.

Bhatt, R., et al., (2008). Synthesis, characterization, and biodegradation of carboxy methyl chitosan-g-medium chain length polyhydroxyalkanoates. *J. Appl. Polym. Sci., 110*, 975–982.

Bordoloi, N., & Konwar, B. K., (2009). Bacterial biosurfactant in enhancing solubility and metabolism of petroleum hydrocarbons. *J. Hazardous Materials, 170*, 495–505.

Bozani, D. K., et al., (2011). Silver nanoparticles encapsulated in glycogen biopolymer: Morphology, optical and antimicrobial properties. *Carbohydrate Polymers, 83*, 883–890.

Budwill, K., et al., (1992). Methanogenic degradation of poly (3-hydroxyalkanoates). *Appl. Environ. Microbiol., 58*, 1398–1401.

Carrasco, F., et al., (2006). Thermal stability of polyhydroxyalkanoates. *J. Appl. Polym. Sci., 100*, 2111–2121.

Colak, A., & Guner, S., (2004). Polyhydroxyalkanoate degrading hydrolase-like activities by *Pseudomonas* sp. isolated from soil. *Int. Biodet. Biodegrad., 53*, 103–109.

Dalal, J., et al., (2010). Evaluation of bacterial strains isolated from oil-contaminated soil for production of polyhydroxyalkanoic acids (PHA). *Pedobiologia, 54*, 25–30.

De-Keijser, T. H., et al., (1982). Use of the Voigt function in a single-line method for the analysis of X-ray diffraction line broadening. *J. Appl. Cryst., 15*, 308–314.

Dutta, S., et al., (2010). Biodegradation of epoxy and MF modified Polyurethane films derived from a sustainable resource. *J. Polym. Environ., 18*, 167–176.

Garland, P. B., (1977). Energy transduction in microbial systems. *Symp. Soc. Gen. Microbiol., 27*, 1–21.

Gunaratne, L. M. W. K., & Shanks, R. A., (2005). Melting and thermal history of poly (hydroxybutyrate-co-hydroxyvalerate) using step-scan DSC. *Thermochim. Acta, 430*, 183–190.

Haba, E., et al., (2007). Poly-3- (hydroxyalkanoate) produced from oily substrates by *Pseudomonas aeruginosa* 47T2 (NCBIM40044): Effect of nutrients and incubation temperature on polymer composition. *Biochem. Eng. J., 35*, 99–106.

He, W., et al., (1998). Production of novel polyhydroxyalkanoates by *Pseudomonas stutzeri* 1317 from glucose and soybean oil. *FEMS Microbiol. Lett., 169*, 45–49.

Huijberts, G. N., et al., (1992). *Pseudomonas putida* KT2442 cultivated on glucose accumulates poly (3-hydroxyalkanoates) consisting of saturated and unsaturated monomers. *Appl. Environ. Microbiol., 58*, 536–544.

Jendrossek, D., & Handrick, R., (2002). Microbial degradation of polyhydroxyalkanoates. *Annu. Rev. Microbiol., 56*, 403–432.

Jendrossek, D., (2002). Extracellular polyhydroxyalkanoate depolymerases: The key enzymes of PHA degradation. In: Doi, Y., & Steinbuchel, A., (eds.), *Biopolymers* (pp. 41–84). Wiley-VCH: Weinheim, Germany.

Kaplan, C. W., & Kitts, C. L., (2004). Bacterial succession in petroleum land treatment unit. *Appl. Environ. Microbiol., 70*, 1777–1786.

Khanna, S., & Srivastava, A. K., (2005). Recent advances in microbial polyhydroxyalkanoates. *Process Biochem., 40*, 607–619.

Kim, D. Y., & Rhee, Y. H., (2003). Biodegradation of microbial and synthetic polyesters by fungi. *Appl. Microbiol. Biotechnol., 61*, 300–308.

Knoll, M., et al., (2009). The PHA depolymerase engineering database: A systematic analysis tool for the diverse family of polyhydroxyalkanoate (PHA) depolymerases. *BMC Bioinformatics* 10, 1–8.

Labuzek, S., & Radecka, I., (2001). Biosynthesis of copolymers of PHB tercopolymers by *Bacillus cereus* UW85strain. *J. Appl. Microbiol., 90*, 353–357.

Lee, S. Y., (1995). Review bacterial polyhydroxyalkanoates. *Biotechnol. Bioeng., 49*, 1–14.

Li, Z. G., et al., (2007). Study of enzymatic degradation of microbial copolyesters consisting of 3-hydroxybutyrate and medium-chain-length 3-hydroxyalkanoates. *Polym. Degrad. Stab., 92*, 1708–1714.

Lim, S. P., et al., (2005). Degradation of medium-chain-length polyhydroxyalkanoates in tropical forest and mangrove soils. *Appl. Biochem. Biotechnol., 126*, 23–33.

Liu, W. K., & Chen, G. Q., (2007). Production and characterization of Mcl-PHA with high 3-hydroxy tetradecanoate monomer content by fadB and fadA knockout mutant of *Pseudomonas putida* KT2442. *Appl. Microbiol. Biotechnol., 76*, 1153–1159.

Loo, C. Y., & Sudesh, K., (2007). Polyhydroxyalkanoates: Bio-based microbial plastics and their properties. *Malaysian Polymer J., 2*(2), 31–57.

Mas-Castella, J., et al., (1995). Biodegradation of poly-β-hydroxyalkanoates in anaerobic sediments. *Int. Biodet. Biodegrade., 35*(1–3), 155–174.

McLafferty, F. W., (1956). Mass spectrometric analysis broad applicability to chemical research. *Anal. Chem., 28*, 306–316.

Misra, A. K., et al., (2000). Screening of poly-hydroxybutyrate-producing microorganisms using Fourier transform infrared spectroscopy. *Biotechnol. Lett., 22*, 1217–1219.

Możejko-Ciesielska, J., & Kiewisz, R., (2016). Bacterial polyhydroxyalkanoates. *Microbiol. Res., 192*, 271–282.

Nishida, H., & Tokiwa, Y., (1993). Effects of higher-order structure of poly (3-hydroxybutyrate) on its biodegradation. II. Effects of crystal structure on microbial degradation. *J. Environ. Polym. Degrad., 1*, 65–80.

Obruca, S., et al., (2010). Production of poly (3-hydroxybutyrate-co-3-hydroxyvalerate) by *Cupriavidusne- cator* from waste rape seed oil using propanol as a precursor of 3-hydroxyvalerate. *Biotechnol. Lett., 32*, 1925–1932.

Ojumu, T. V., et al., (2004). Production of polyhydroxyalkanoates, a bacterial biodegradable polymer. *African J. Biotech., 3*(1), 18–24.

Ostle, A. G., & Holt, J. G., (1982). Nile Blue A as a fluorescent stain for poly-b-hydroxybutyrate. *Appl. Environ. Microbiol., 44*, 238–241.

Pederson, E. N., et al., (2006). Bacterial synthesis of PHA block copolymer. *Biomacromol., 6*, 1904–1911.

Poirier, Y., et al., (1995). Production of polyhydroxyalkanoates, a family of Biodegradable plastics and elastomers, in bacterial and plant. *Biotechnol., 13*, 142–150.

Potter, M., & Steinbuchel, A., (2005). Poly (3-hydroxybutyrate) granule-associated proteins: Impacts on poly (3-hydroxybutyrate) synthesis and degradation. *Biomacromol., 6*, 552–560.

Rai, R., et al., (2011). Medium chain length polyhydroxyalkanoates, promising new biomedical materials for the future. *Mat. Sci. Engg. R Reports, 72*(3), 29–47.

Rehman, S., et al., (2007). Screening of different contaminated environments for polyhydroxyalkanoates-producing bacterial strains. *Biol. Bratislava, 62*, 650–656.

Sanchez, R., et al., (2003). Medium-chain-length polyhydroxyalkanoic acids (PHA Mcl) produced by *Pseudomonas putida* IPT 046 from renewable sources. *Eur. Polym J., 39*, 1385–1394.

Sato, H., et al., (2005). Compositional analysis of poly (3-hydroxybutyrate-co-3-hydroxyvalerate) by pyrolysis-gas chromatography in the presence of organic alkali. *J. Anal. Appl Pyrolysis, 74*, 193–199.

Savenkova, L., et al., (2000). Mechanical properties and biodegradation characteristics of PHB-based films. *Process Biochem., 35*, 573–579.

Shah, A. A., et al., (2010). Degradation of poly (3-hydroxybutyrate-co-3-hydroxyvalerate) by a newly isolated *Actinomadura* sp. AF-555, from soil. *Int. Biodet. Biodegrad., 64*, 281–285.

Shangguan, Y. Y., et al., (2006). The mechanical properties and *in vitro* biodegradation and biocompatibility of UV-treated poly (3-hydroxybutyrate-co-3-hydroxyhexanoate). *Biomaterials, 27*(11), 2349–2357.

Singh, V. R., (2015). Polyhydroxybutyrate (PHB): Biodegradable bioplastics produced by microorganisms. *Int. Jf. Pharma. Res., 7*(2), 17–23.

Solaiman, D K Y., et al., (2002). Physiological characterization and genetic engineering of *Pseudomonas corrugata* for medium-chain-length polyhydroxyalkanoates synthesis from triacylglycerols. *Curr. Microbiol., 44*, 189–195.

Sridewi, N., et al., (2006). Degradation of commercially important polyhydroxyalkanoates in tropical mangrove ecosystem. *Polym. Degrad. Stab., 91*, 2931–2940.

Thire, R. M. S. M., et al., (2006). Effect of starch addition on compression-molded Poly (3-hydroxy-butyrate)/starch blends. *J. Appl. Polym. Sci., 100*, 4338–4347.

Tobin, K. M., et al., (2007). Polyphosphate accumulation by *Pseudomonas putida* CA-3 and other medium-chain-length polyhydroxyalkanoate accumulating bacteria under aerobic growth condition. *Appl. Environ. Microbiol., 73*, 1383–1387.

Tsuge, T., (2002). Metabolic improvements and use of inexpensive carbon sources in microbial production of polyhydroxyalkanoates. *J. Biosci Bioeng., 94*, 579–584.

Valappil, S. P., et al., (2007). Polyhydroxyalkanoate (PHA) biosynthesis from structurally unrelated carbon sources by a newly characterized *Bacillus* spp. *J. Biotech, 127*(3), 475–487.

Van De, V. K., & Kiekens, P., (2002). Biopolymers: Overview of several properties and consequences on their applications. *Polymer Testing, 21*(4), 433–442.

Verlinden, R. A. J., et al., (2011). Production of polyhydroxyalkanoates from waste frying oil by *Cupriavidus necator*. *AMB Express, 1*, 1–8.

Vijay, R., & Tarika, K., (2018). Production of polyhydroxyalkanoates (PHAs) using synthetic biology and metabolic engineering approaches. *Res. J. Biotech, 13*(1), 99–109.

Wang, Y. W., et al., (2004). Biodegradation studies of poly (3-hydroxybutyrate-co-3-hydroxyhexanoate). *Polym. Degrad. Stab., 85*, 815–821.

William, J. P., & Christopher, J. T., (1996). Quantitation of poly-β-hydroxybutyrate by fluorescence of bacteria and granules stained with Nile blue A. *Biotechnol. Tech., 10*, 215–220.

Wong, A. L., et al., (2000). Microbial production of polyhydroxyalkanoates by bacteria isolated from oil wastes. *Appl. Biochem. Biotechnol., 84*, 843–857.

Zinn, M., et al., (2001). Occurrence, synthesis and medical application of bacterial polyhydroxyalkanoates. *Adv. Drug Delivery Rev., 53*, 5–21.

CHAPTER 10

Biochemical Characterization of the PHA-Producing Bacterial Strains

The PHA-producing bacteria were characterized on the basis of biochemical tests. All three bacterial strains were assessed by following the standard protocol. Data thus obtained are presented in Table 10.1. Characters recorded for the tests were catalase, urease, citrate, triple sugar iron, nitrate reduction, indole, and H_2S production, litmus milk reaction, MR-VP, starch, gelatin, and casein hydrolysis, glucose, and combination of $(NH_4)_2SO_4$ and glucose utilization and acid production from sugars like lactose, mannitol, dextrose, sucrose, and xylose.

TABLE 10.1 Biochemical Characterization of the PHA Producing Bacterial Strains

Enzymes	**Biochemical Characters**		
	P. aeruginosa		***B. circulans* MTCC8167**
	JQ796859	**JQ866912**	
Catalase	+	+	+
Urease	–	–	+
Citrate	+	+	–
Triple sugar iron	+	+	+
Nitrate reduction	+	+	–
Indole production	–	–	–
H_2S production	–	–	–
Litmus milk reaction	Peptonization	Peptonization	Peptonization
Methyl red	–	+	–
Voges Proskauer	–	–	–

Bacterial Biopolymers. B. K. Konwar, PhD (Author)

TABLE 10.1 *(Continued)*

Enzymes	**Biochemical Characters**		
	P. aeruginosa		***B. circulans* MTCC8167**
	JQ796859	**JQ866912**	
Starch hydrolysis	–	–	+
Gelatin hydrolysis	+	+	+
Casein hydrolysis	+	+	+
1% Glucose	+	+	+
$(NH_4)_2SO_4$+ Glucose	+	+	+
Acid Production from Sugars:			
Lactose	+	–	+
Mannitol	–	+	+
Dextrose	+	+	+
Sucrose	+	+	+
Xylose	+	+	+

10.1 GROWTH OF PHA PRODUCING BACTERIAL STRAINS AT DIFFERENT PH LEVELS

The PHA producing bacterial strains were assessed for their growth in the different pH levels from acidic to alkaline (pH 2–10) media containing glucose. Data thus obtained are presented in Figure 10.1. All the bacterial strains showed poor growth at the high acidic (pH 2) and high alkaline (pH 10) conditions. At pH 2 the bacterial strains *P. aeruginosa* JQ796859, *B. circulans* MTCC8167 and *P. aeruginosa* JQ866912 produced comparatively lower biomass of 0.25–0.34 gL^{-1} having the peak at 96 h. The strains *B. circulans* MTCC8167 and *P. aeruginosa* JQ866912 retarded their growth after 96 h, but *P. aeruginosa* JQ796859 showed stationary growth up to 120 h. At pH 4, all strains showed similar pattern of growth, however, *P. aeruginosa* JQ796859 showed better growth in comparison to the others. In the case of pH 6, *B. circulans* MTCC8167 exhibited better growth with biomass yield of 1–1.48 gL^{-1} during the initial culture period of 24–72 h in comparison to the other two strains. However, a further increase in pH (up to pH 10) resulted in the decline in growth for all strains.

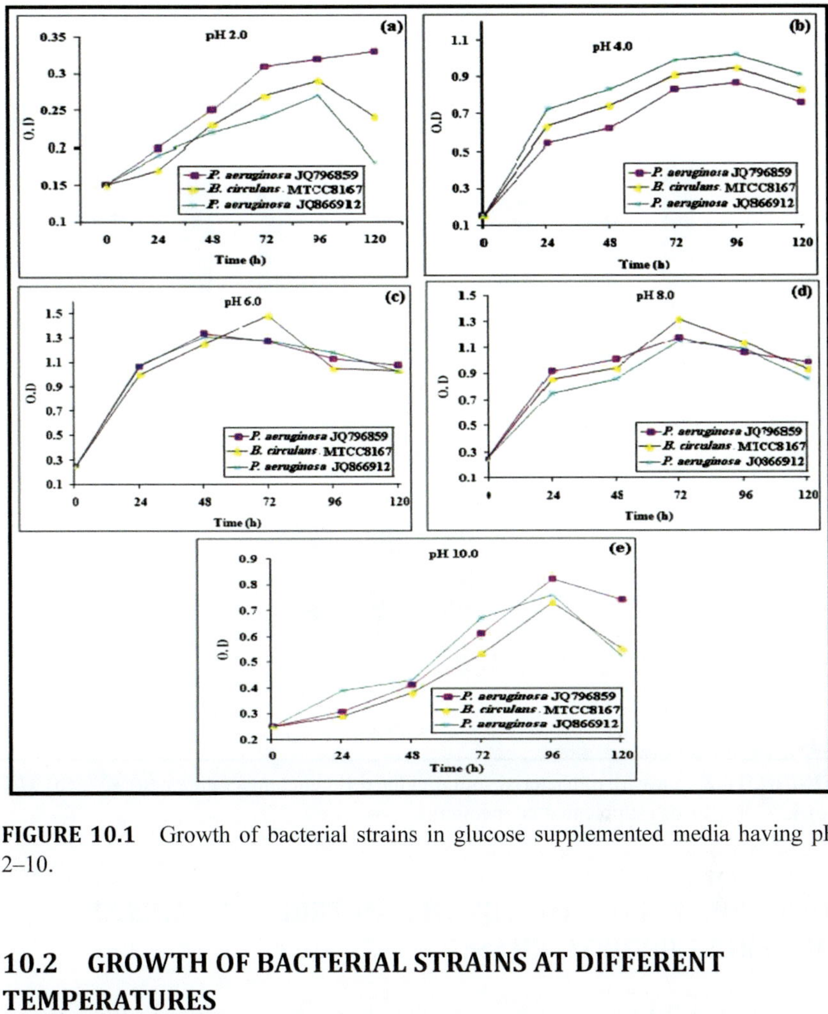

FIGURE 10.1 Growth of bacterial strains in glucose supplemented media having pH 2–10.

10.2 GROWTH OF BACTERIAL STRAINS AT DIFFERENT TEMPERATURES

The PHA producing bacterial strains were assessed for their growth at the temperature range of 30–45°C in the medium containing glucose. Data thus obtained are presented in Figure 10.2. The effect of temperature on the growth of the bacterial strains showed that the yield of biomass was significantly increased at temperature 30–37°C. At 37°C, the strains showed good biomass production; *P. aeruginosa* JQ866912 exhibited the highest biomass production of 4.3 gL^{-1} as compared to the others. However, all the bacterial strains failed to survive beyond this temperature and the yield of biomass also declined. Therefore, 37C was considered as the optimum temperature for the growth of the bacterial strains.

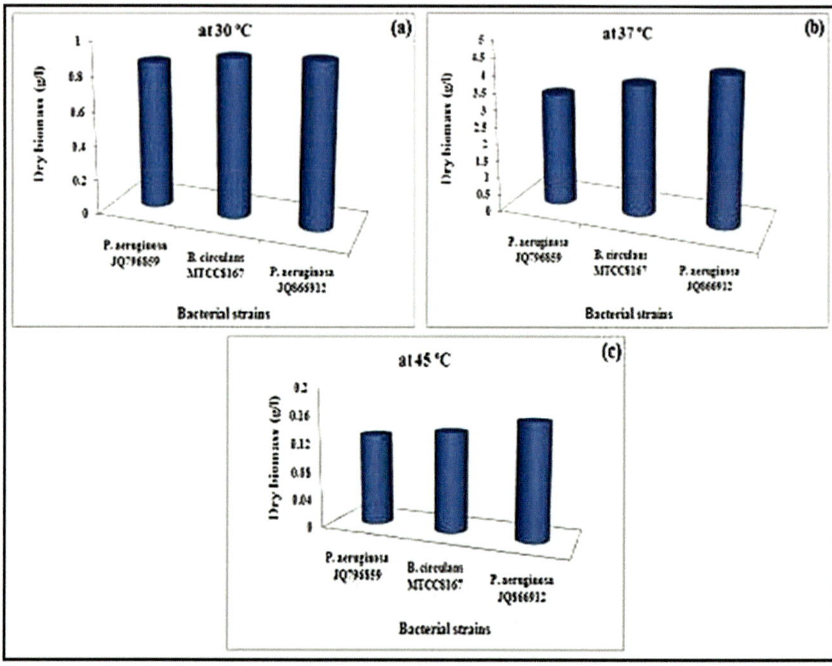

FIGURE 10.2 Growth of bacterial strains at the different temperatures: (a) 30°C (b) 37°C (c) 45°C in Glucose supplemented medium.

10.3 PHA PRODUCTION BY THE BACTERIAL STRAINS AT DIFFERENT GROWTH PHASES

The selected PHA-producing bacterial strains were grown in PHA detection medium containing glucose (1%) for the different time periods in the production and accumulation of PHA. The accumulated PHA was assayed at the time interval of 12 h till 96 h. Data thus obtained are presented in Table 10.2.

TABLE 10.2 PHA Production by the Bacterial Strains at Different Growth Periods

Isolate	PHA (% CDW) in Glucose Supplemented Medium				
	12 h	24 h	48 h	72 h	96 h
P. aeruginosa JQ796859	ND	1.4	4.5	6.47	5.1
B. circulans MTCC8167	ND	1.9	5.6	9.22	6.9
P. aeruginosa JQ866912	ND	2.5	10.4	18.0	16.2

Note: ND: Not detectable; the values presented here are the mean of three independent experiments.

From the table, it was observed that during the first 12 h period there was no accumulation of PHA in all three bacterial strains. At 24 h culture, the strains showed initiation of PHA accumulation. Then, the accumulation of PHA gradually increased up to 72 h of culture. The highest PHA accumulation was observed in *P. aeruginosa* JQ866912, 18.0% after 48 to 72 h of culture. Other two strains also showed increased accumulation of PHA (*P. aeruginosa* JQ796859, 6.5% and *B. circulans* MTCC8167, 9.2%) after 48 to 72 h of culture. But all three bacterial cultures exhibited reduction in the accumulated PHA after 96 h of culture (*P. aeruginosa* JQ796859: 5.1%, *B. circulans* MTCC8167: 6.9% and *P. aeruginosa* JQ866912: 16.2%).

10.4 PHA PRODUCTION BY THE BACTERIAL STRAINS IN DIFFERENT CARBON SUBSTRATES

Various carbon sources were screened for the production of biopolymer by the bacterial strains. Biopolymer producing bacterial strains was separately inoculated in media supplemented with the different carbon sources: glucose, glycerol, Colocasia starch, propionic acid, soy oil, waste mobile, sugarcane molasses and glycerol byproduct of kitchen chimney dump lard (KCDL). The growth of the bacterial isolates and PHA yield are presented in Table 10.3.

TABLE 10.3 PHA Production by the Bacterial Isolates Using the Different Carbon Sources

Isolate	Carbon Source	Biomass (g.L^{-1})	PHA (% CDW)
P. aeruginosa JQ796859	Glucose (1%)	5.4	6.47
B. circulans MTCC8167		6.32	9.22
P. aeruginosa JQ866912		6.8	18.0
P. aeruginosa JQ796859	Glycerol (1%)	4.5	5.6
B. circulans MTCC8167		5.9	8.0
P. aeruginosa JQ866912		7.4	14.5
P. aeruginosa JQ796859	Colocasia starch (1%)	0.6	–
B. circulans MTCC8167		4.8	6.5
P. aeruginosa JQ866912		0.3	–
P. aeruginosa JQ796859	Glucose (1%) + Propionic acid (0.1%)	3.6	4.6
B. circulans MTCC8167		6.8	19.8
P. aeruginosa JQ866912		5.5	12.5
P. aeruginosa JQ796859	Sugarcane molasses (1%)	4.1	4.7
B. circulans MTCC8167		5.3	7.9
P. aeruginosa JQ866912		6.3	13.2
P. aeruginosa JQ796859	Waste mobile (1%)	2.3	1.2
B. circulans MTCC8167		–	–
P. aeruginosa JQ866912		2.8	1.4
P. aeruginosa JQ796859	Soy oil (1%)	5.9	7.77
B. circulans MTCC8167		6.2	8.12
P. aeruginosa JQ866912		7.6	17.07
P. aeruginosa JQ796859	Glycerol waste of KCDL (1%)	5.7	7.9
B. circulans MTCC8167		4.9	6.3
P. aeruginosa JQ866912		7.8	22.5

NB: The values presented here are the mean of three independent experiments.

All the bacterial strains were tested for their PHA accumulation against the different carbon sources (1%) in the medium after 72 h of culture. In the glucose containing medium the isolate *P. aeruginosa* JQ866912 exhibited the highest PHA accumulation of 18.0% CDW followed by *B. circulans* MTCC8167 and *P. aeruginosa* JQ796859. Similar results were obtained in the glycerol containing medium; the highest 14.5% PHA accumulation was shown by *P. aeruginosa* JQ866912 isolate.

But, in the case of Colocasia starch containing medium, no growth was exhibited by the strains *P. aeruginosa* JQ796859 and JQ866912; but *B. circulans* MTCC8167 showed 6.5% accumulation. The strain *B. circulans* MTCC8167 exhibited the highest PHA accumulation of 19.8% in the medium supplemented with the combination of 0.1% propionic acid and 1% glucose followed by *P. aeruginosa* JQ866912 (12.5%) and *P. aeruginosa* JQ796859 (4.6%). Almost, the similar result was exhibited by *P. aeruginosa* JQ796859 in sugarcane molasses as the carbon source, as well as glucose and propionic acid combination. The strain *P. aeruginosa* JQ866912 accumulated 13.2% PHA in the same source. While using waste mobile as the carbon source in the medium, all the strains showed the least accumulation of PHA. All the strains showed better growth and PHA accumulation in soy oil-supplemented medium. The strain *P. aeruginosa* JQ866912 exhibited 17.07% PHA followed by *B. circulans* MTCC8167 and *P. aeruginosa* JQ796859. Comparatively, the best PHA accumulation was observed when the glycerol byproduct (waste of KCDL) was used; the strain *P. aeruginosa* JQ866912 showed 22.5% CDW of PHA accumulation followed by *P. aeruginosa* JQ796859 and *B. circulans* MTCC8167 of 7.9–6.3% PHA.

The bacterial biomass yield in shaking flask cultures with the use of the above carbon sources as the sole source of carbon (1% w/v) are shown in Table 10.4. After 72 h of culture, each 5 mL sample was assayed for dry biomass yield. The results revealed that higher cell densities were obtained from glycerol byproduct. Cell growth up to a dry biomass of 7.8 gL^{-1} from the bacterial strain *P. aeruginosa* JQ866912 was obtained while using the glycerol byproduct supplemented medium. In the case of medium with Colocasia starch and waste mobile there was a decrease in the cell biomass yield in both *P. aeruginosa* JQ796859 and *P. aeruginosa* JQ866912, but *B. circulans* MTCC8167 showed an increase up to 4.8 $g.l^{-1}$.

10.5 LIGHT ABSORPTION BY PHA

The isolated PHA was dissolved in chloroform and subjected to the wavelength scan to determine the light absorption at 190–600 nm. The maximum absorption (λ_{max}) in the wavelength was thus recorded by using UV-Vis Spectrophotometer. Data thus obtained are presented graphically in Figure 10.3(a–c). PHA isolated from *P. aeruginosa* JQ796859 was found to have its maximum absorption at the wavelength of 261 nm; PHA from *B. circulans* MTCC8167 showed at 289 nm and PHA from *P. aeruginosa* JQ866912 showed maximum absorption at 285–290 nm.

10.6 FTIR ANALYSIS OF PHA

Characterization of the extracted and purified PHA was carried out with the help of Fourier transform infrared spectrometer (FTIR) technique. The FTIR analysis of the isolated polymer of three strains *P. aeruginosa* JQ796859, *B. circulans* MTCC8167 and *P. aeruginosa* JQ866912 revealed absorption bands in 1,730–1,735 cm^{-1}, corresponding to the ester carbonyl group. Other characteristic bands were observed in the spectrum, albeit at a different position and with weaker bands. The FTIR spectral pattern of the polymer compounds of three different bacterial strains were depicted in Figures 10.4–10.7, and Table 10.4. The data were also compared with the control PHB.

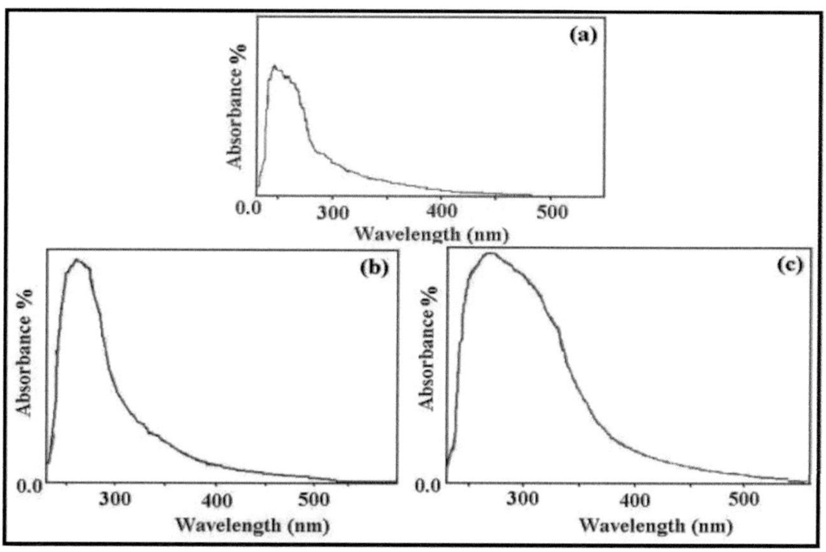

FIGURE 10.3 UV-visible absorption spectrum of PHA from the bacterial strains: (a) *P. aeruginosa* JQ796859; (b) *B. circulans* MTCC8167; and (c) *P. aeruginosa* JQ866912.

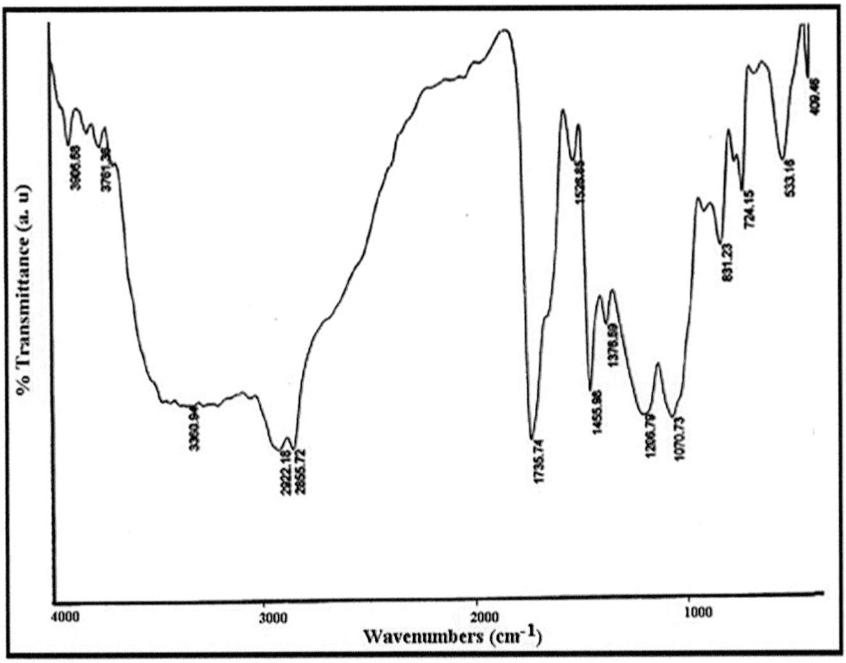

FIGURE 10.4 FTIR spectra of purified polymer from the bacterial strain *P. aeruginosa* JQ796859 when cultured in glucose containing media.

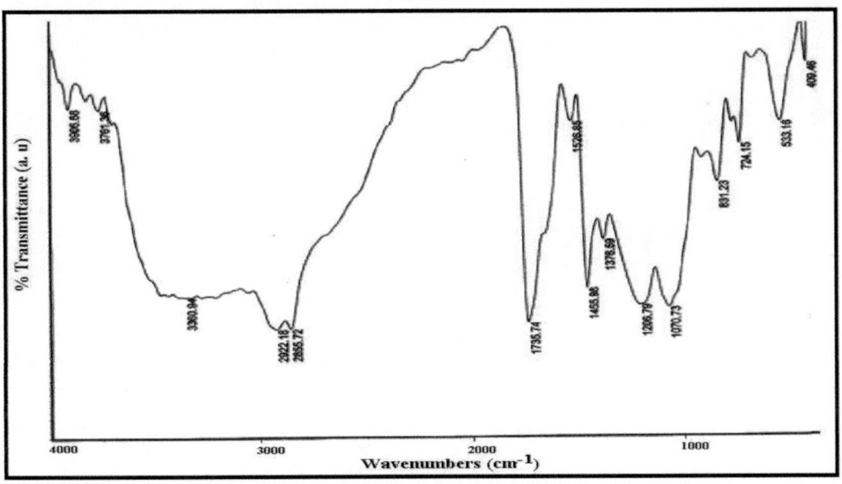

FIGURE 10.5 FTIR spectra of purified polymer from the bacterial strain *B. circulans* MTCC8167 when cultured in glucose containing media.

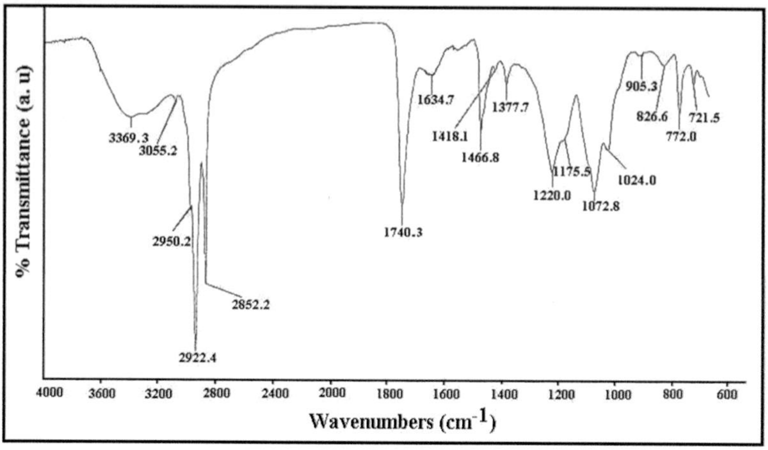

FIGURE 10.6 FTIR spectra of purified polymer from the bacterial strain *P. aeruginosa* JQ866912 when cultured in glucose containing media.

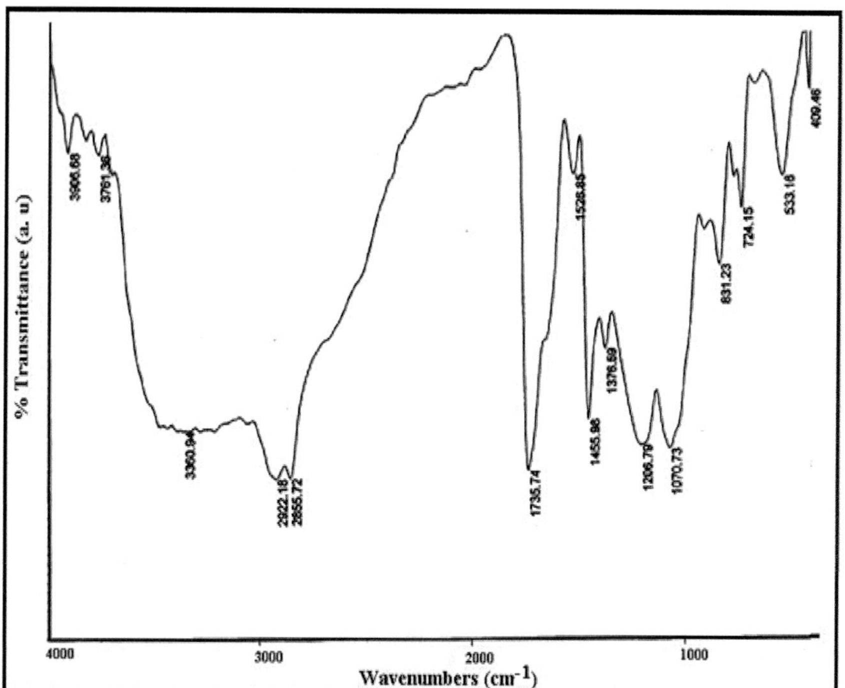

FIGURE 10.7 FTIR spectra of purified polymer from the bacterial strain *B. circulans* MTCC8167 when cultured in glucose containing media.

TABLE 10.4 Structural Information Generated from FTIR Data

Bacterial Strain	FTIR Peaks (Wave No. cm^{-1})	Structural Information Revealed
P. aeruginosa JQ796859	3,405	O–H bonding
	2,923	Bands of C–H stretching
	1,730	Band of ester carbonyl C = O
	1,670	–C = C stretching
	3,085	Alkene C–H stretching
	1,065	–C–O–C stretching vibrations
B. circulans MTCC8167	3,360	O–H bonding
	2,922	Bands of C–H stretching
	1,735	Strong absorption band of ester carbonyl C = O
	1,206	C–O–C group
	1,070	–C–O–C stretching vibrations
P. aeruginosa JQ866912	3,369	O–H bonding
	2,922	Bands of C–H stretching
	1,740	Strong absorption band of ester carbonyl C = O
	1,634	–C = C
	3,055	Alkene C–H stretching
	1,071	–C–O–C stretching vibrations

The FTIR peaks were identical with the data provided by different researchers (Kadouri et al., 2005; Berlanga et al., 2006; Williamson and Wilkinson, 1958). The comparisons suggested that the polymer compounds might be polyhydroxyalkanoates (PHAs).

10.7 GC-MS ANALYSIS OF POLYHYDROXYALKANOATES (PHAS)

The actual monomeric characterization of the biopolymer obtained from the bacterial strains was done using GC-MS. For the analysis, both the dry cell sample and the isolated polymer were methanolized to convert the polymer into methyl-3-hydroxyalkanoate with the concerned functional group.

10.7.1 PHA FROM BACTERIAL STRAIN P. AERUGINOSA JQ796859

In the case of the polymer from *P. aeruginosa* JQ796859 biopolymer, the GC pattern of the total ion chromatogram (TIC) possessed two methyl ester peaks. To identify these two major peaks, mass spectroscopy measurement was performed and compared with the mass spectrum of methyl esters in the MS library as shown in Figures 10.8 and 10.9.

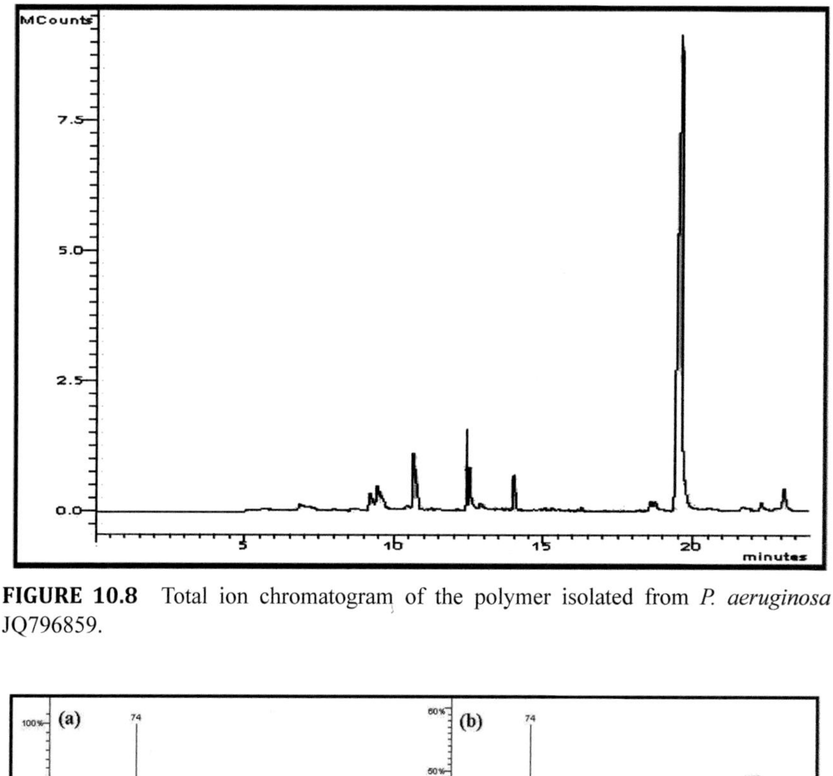

FIGURE 10.8 Total ion chromatogram of the polymer isolated from *P. aeruginosa* JQ796859.

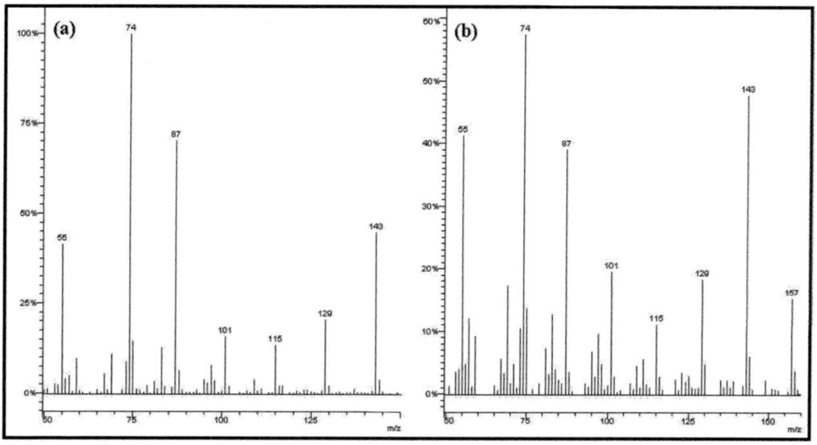

FIGURE 10.9 Mass spectrums of methyl ester peaks with the retention time RT = 11.45 (a); and RT = 19.65, (b) min of the polymer isolated from *P. aeruginosa* JQ796859.

10.7.2 PHA FROM BACTERIAL STRAIN *B. CIRCULANS* MTCC8167

The TIC for the methanolysis product of the isolated PHA from the bacterial isolate *B. circulans* MTCC8167 possessed two methyl ester peaks (Figure 10.10). To identify these two major peaks, mass spectroscopy measurement was performed and found similarity with the mass spectrum of methyl esters in the MS library as shown in Figure 10.11(a) and (b).

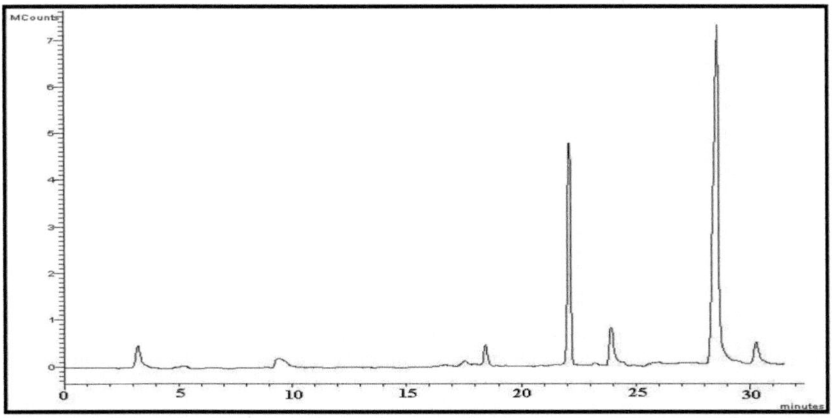

FIGURE 10.10 Total ion chromatogram of the polymer isolated from *B. circulans* MTCC8167.

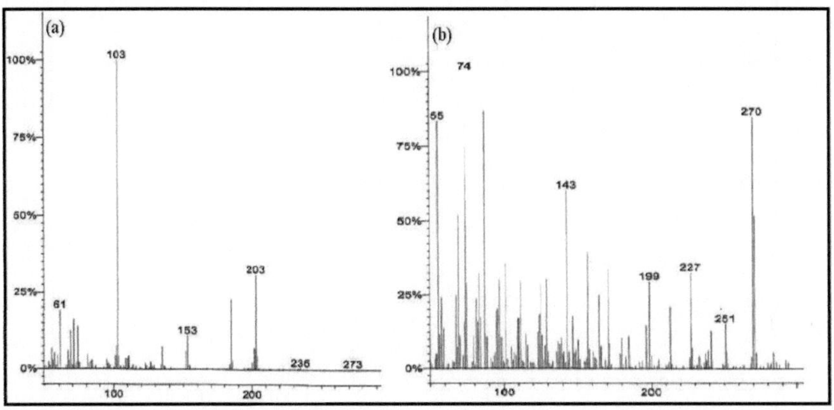

FIGURE 10.11 Mass spectra of methyl ester peaks with the retention time RT = 22.67 (a); and RT = 28.10 (b) min of the polymer isolated from *B. circulans* MTCC8167.

10.7.3 PHA FROM BACTERIAL STRAIN P. AERUGINOSA JQ866912

In the case of the polymer from *P. aeruginosa* JQ866912, the GC pattern of the TIC (Figure 10.12) possessed three methyl ester peaks. To identify these three major peaks, mass spectroscopy measurement was performed and compared with the mass spectrum of methyl ester in the MS library as shown in Figure 13(a–c).

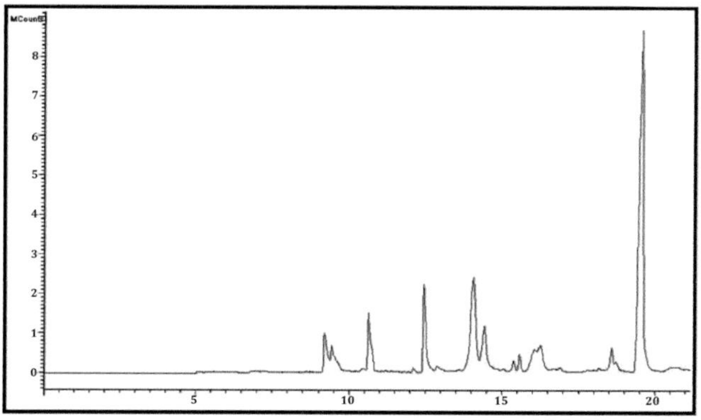

FIGURE 10.12 Total ion chromatogram of the biopolymer isolated from *P. aeruginosa* JQ866912.

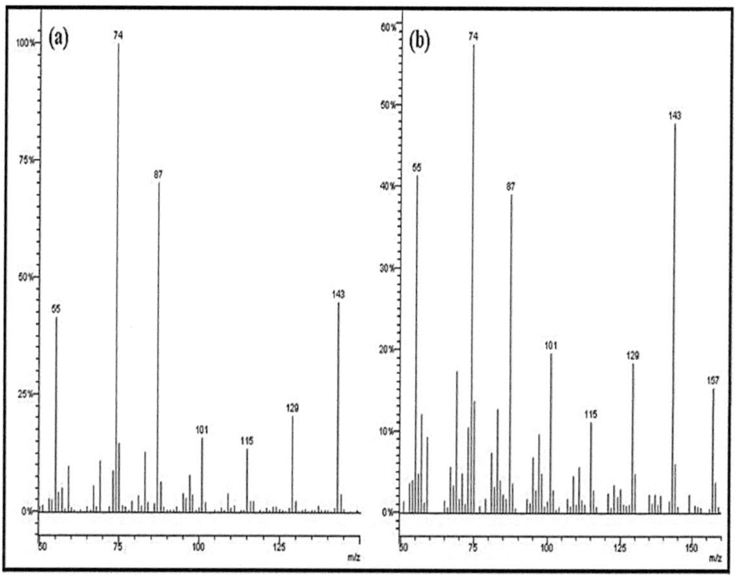

FIGURE 10.13 Mass spectrums of methyl ester peaks with the retention time RT = 11.56 (a); RT = 18.46 (b); and RT = 19.66 (c) min of the polymer isolated from *P. aeruginosa* JQ866912 strain.

10.8 NMR ANALYSIS OF POLYHYDROXYALKANOATES (PHAS)

To confirm the chemical structure of PHA, ^{1}H and ^{13}C NMR analyzes were used. The spectra were measured with JEOL JNN-ECS 400 spectrophotometer. ^{1}H NMR (400 MHz) and ^{13}C NMR (400 MHz) spectra were recorded at 25°C in a deuterated chloroform ($CDCl_3$) solution. Chemical shifts were reported in ppm relative to the signal of tetramethylsilane. The ^{1}H NMR and ^{13}C NMR spectra of the polymer sample from *P. aeruginosa* JQ796859, *B. circulans* MTCC8167 and *P. aeruginosa* JQ866912 are presented in Figures 10.14(a, b)–10.16(a, b), respectively. The characteristic peaks from the figure revealed that all the strains produced copolymer.

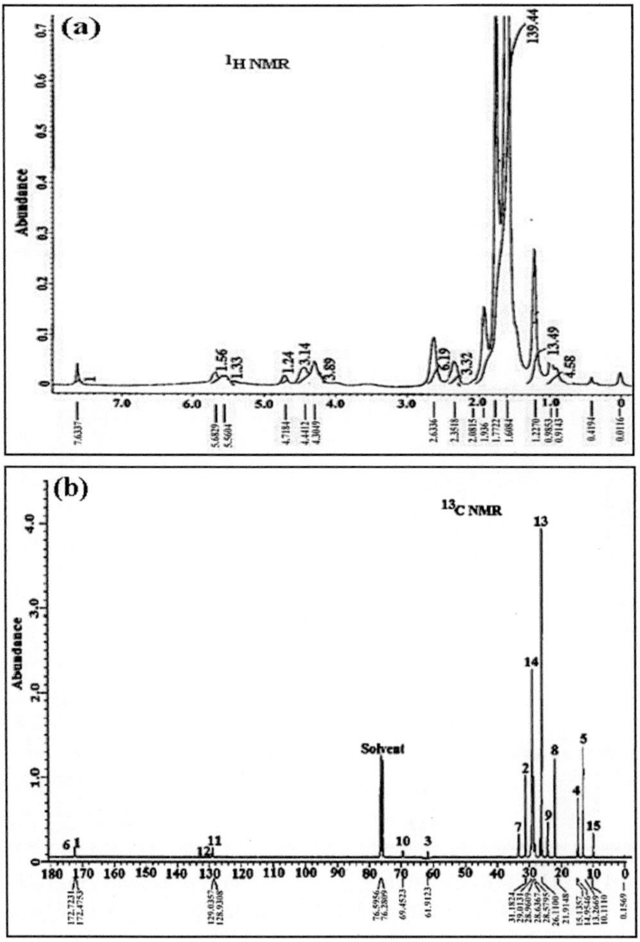

FIGURE 10.14 ^{1}H NMR spectrum (a); and ^{13}C NMR spectrum (b) of P-3HV-5-HDE polymer produced by *P. aeruginosa* JQ796859 when grown on glucose as a carbon substrate.

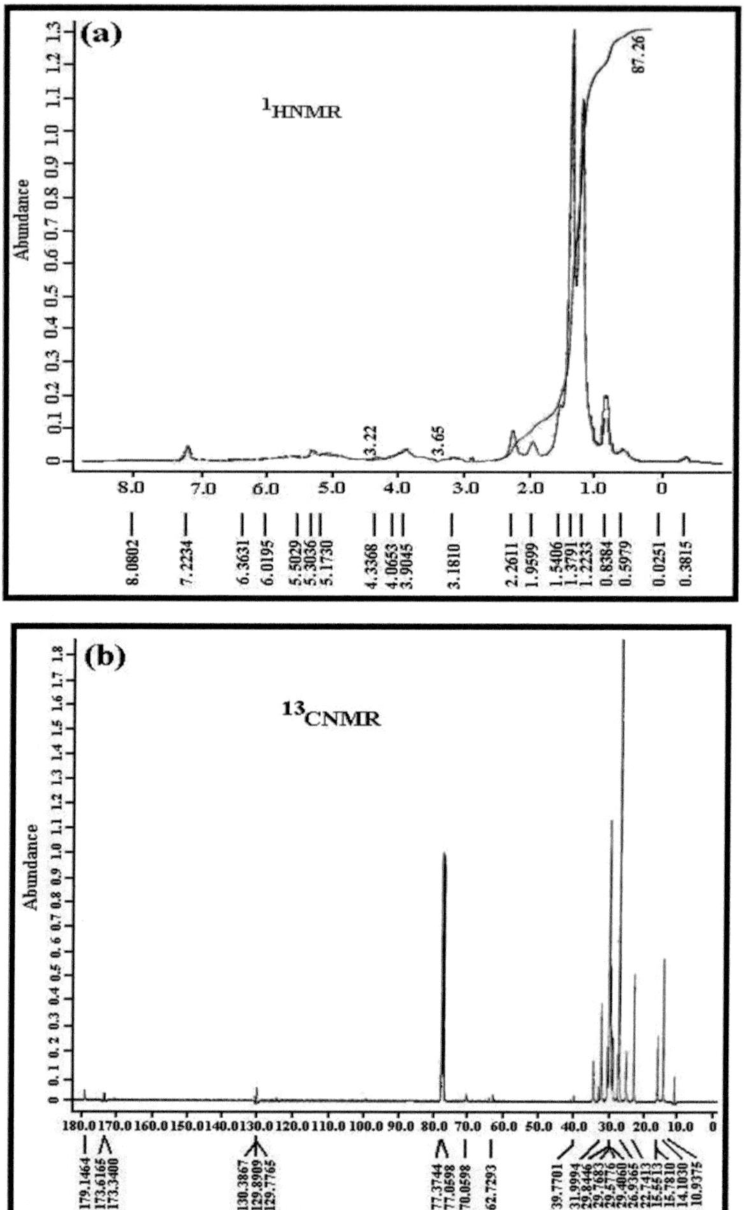

FIGURE 10.15 ^{1}H NMR spectrum (a); and ^{13}C NMR spectrum (b); of P-3HB-3-HV polymer produced by *B. circulans* MTCC8167 when grown on glucose as a carbon substrate.

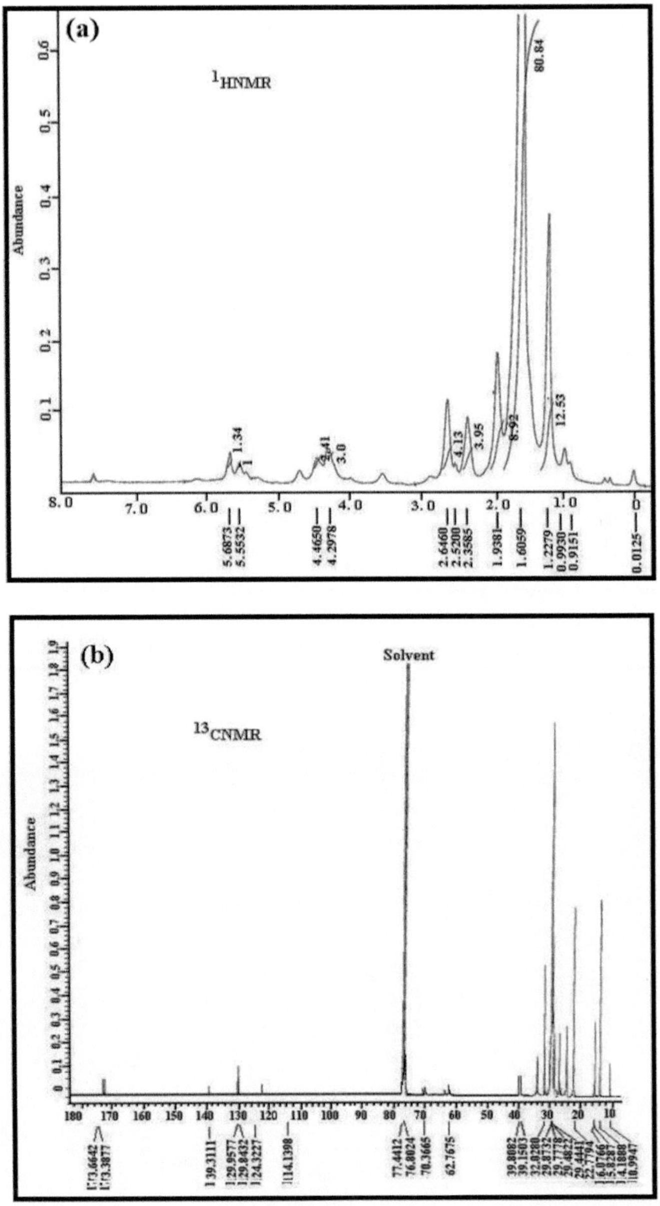

FIGURE 10.16 ^1H NMR spectrum (a); and ^{13}C NMR spectrum (b) of P-3HV-5HDE-3HODE polymer produced by *P. aeruginosa* JQ866912 when grown on glucose as a carbon substrate.

10.9 DETERMINATION OF MOLECULAR WEIGHT (MW) OF PHA GEL PERMEATION CHROMATOGRAPHY (GPC)

The molecular weight (MW) of biopolymer extracted from the bacterial strains *P. aeruginosa* JQ796859, *B. circulans* MTCC8167, and *P. aeruginosa* JQ866912 was measured by using GPC in THF solution using polystyrene standard. Average MWs of the polymers in THF are shown in Table 10.5. GPC curve of the polymers is presented in Figures 10.17–10.19.

TABLE 10.5 Average Molecular Weight of All Three Biopolymers

Polymer Sample	$\bar{M}_n$	$\bar{M}_w$	Polydispersity Index
P. aeruginosa JQ796859	5.6×10^3	5.9×10^3	1.05
B. circulans MTCC8167	4.2×10^4	5.1×10^4	1.21
P. aeruginosa JQ866912	3.8×10^4	4.1×10^4	1.08

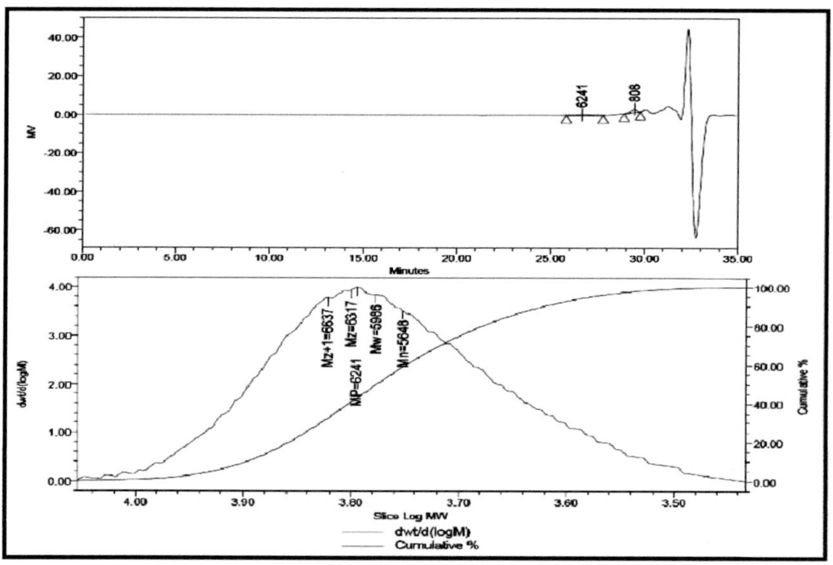

FIGURE 10.17 GPC of *P. aeruginosa* JQ796859 biopolymer.

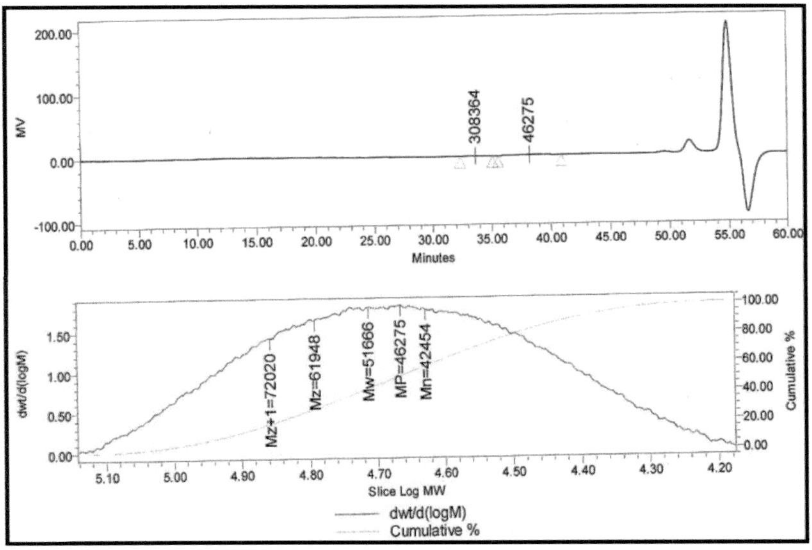

FIGURE 10.18 GPC of *B. circulans* MTCC8167 biopolymer.

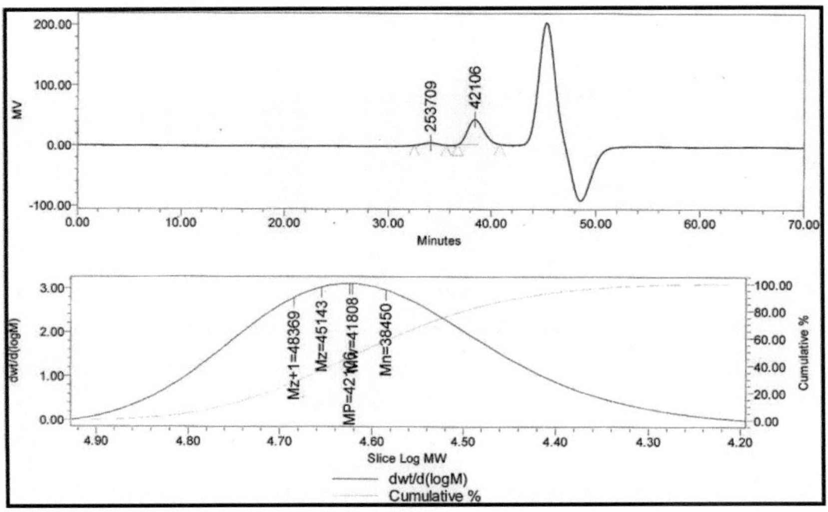

FIGURE 10.19 GPC of *P. aeruginosa* JQ866912 biopolymer.

10.10 CRYSTALLINITY OF PHA BY X-RAY DIFFRACTION (XRD)

X-ray diffraction (XRD) analysis was performed on PHA polymers isolated from *P. aeruginosa* JQ796859, *B. circulans* MTCC8167 and *P. aeruginosa* JQ866912. The calculation on crystallinity of the polymer samples is also presented in Figure 10.20. The XRD data revealed that the polymers are crystalline in nature with large crystal size.

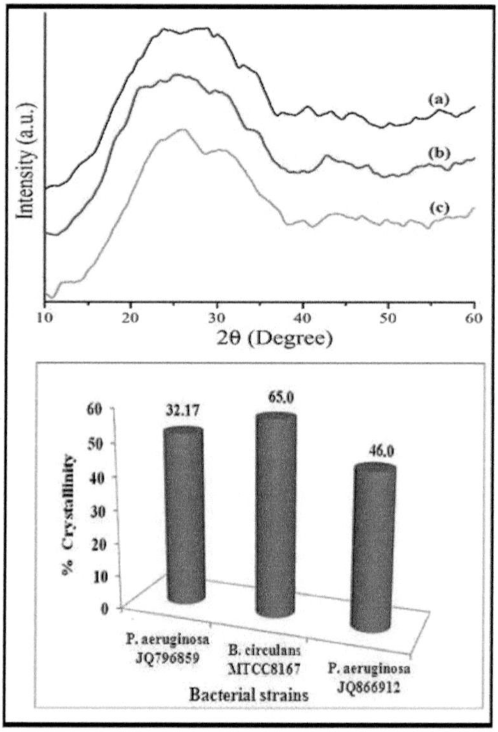

FIGURE 10.20 XRD result of polymer from *P. aeruginosa* JQ796859 (a); *B. circulans* MTCC8167 (b); and *P. aeruginosa* JQ866912 (c) bacterial isolates.

10.11 THERMOGRAVIMETRIC (TGA) ANALYSIS OF PHA BIOPOLYMER

The TGA of the polymers from the bacterial strains *P. aeruginosa* JQ796859, *B. circulans* MTCC8167 and *P. aeruginosa* JQ866912 is presented in Figure 10.21. The figure revealed the thermal degradation profile of all the polymers in comparison to the standard polymer obtained from the market. From Figure 10.22, the percentage of weight loss of the polymer samples was calculated. The biopolymers isolated from the strains showed a high degree of thermal stability in comparison to the standard polymer.

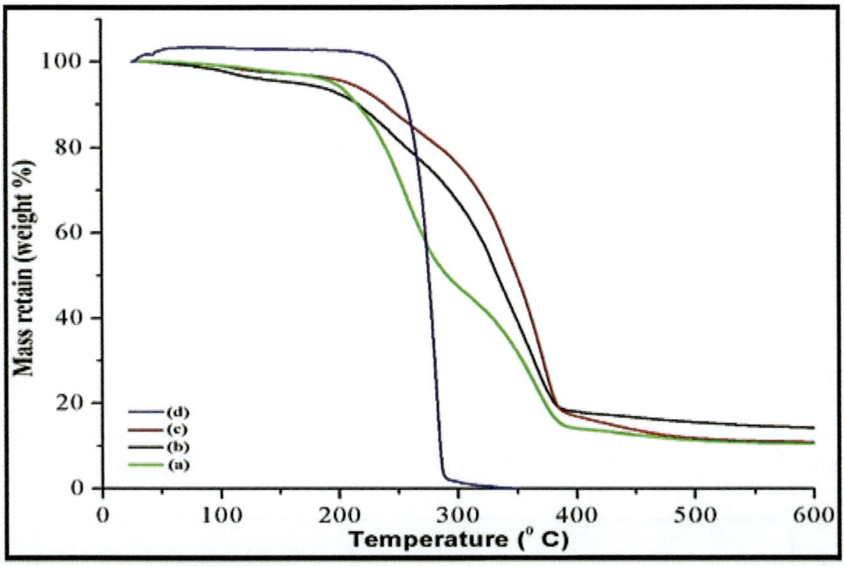

Figure 10.21. Thermogravimetric analysis of the polymers from the strains *P. aeruginosa* JQ796859 (a); *B. circulans* MTCC8167 (b); *P. aeruginosa* JQ866912 (c); and standard polymer (d).

10.12 DIFFERENTIAL SCANNING CALORIMETRY (DSC) ANALYSIS OF PHA

The DSC of the polymer from the bacterial strains *P. aeruginosa* JQ796859, *B. circulans* MTCC8167 and *P. aeruginosa* JQ866912 is presented in Figure 10.22(a–c). The figure depicts changes in the degradation pattern of the polymer samples. All the polymers also have high thermal stability.

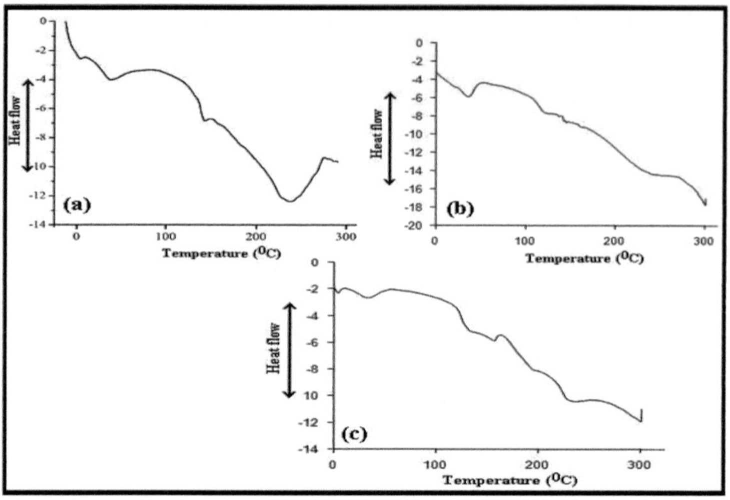

FIGURE 10.22 DSC thermograph analysis of the polymer from *P. aeruginosa* JQ796859 (a); *B. circulans* MTCC8167 (b); and *P. aeruginosa* JQ866912 (c).

10.13 PHOTOLUMINESCENCE (PL) STUDY OF PHA BIOPOLYMERS

The PL emission spectra were recorded at two different wavelengths for all three polymers (Figure 10.23). The PL of the polymers excited at their

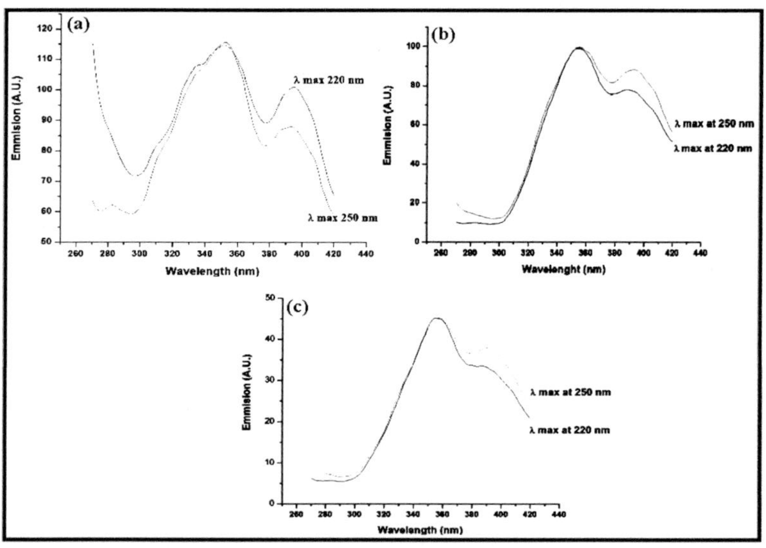

FIGURE 10.23 PL intensity of the polymers isolated from *P. aeruginosa* JQ796859 (a); *B. circulans* MTCC8167 (b); and *P. aeruginosa* JQ866912 (c).

maximum absorption wavelength is measured. All the polymers showed luminescence property. The polymer shows luminescence in the range of 300–420 nm with high intensity.

10.14 SURFACE MORPHOLOGY STUDY OF PHA BIOPOLYMER USING SEM

The surface morphology of the extracted polymer samples of all three bacterial strains was studied by using SEM. For the analysis, a thin film was prepared on to a thin glass slide by carefully spreading PHA solution in chloroform. After that, the slide was left undisturbed for over 24 h for complete drying. The SEM micrograph of the polymer from *P. aeruginosa* JQ796859, *B. circulans* MTCC8167 and *P. aeruginosa* JQ866912 strains are presented in Figure 10.24(a–c). SEM of the polymer samples showed that polymers possessed uneven surface with non-porous texture.

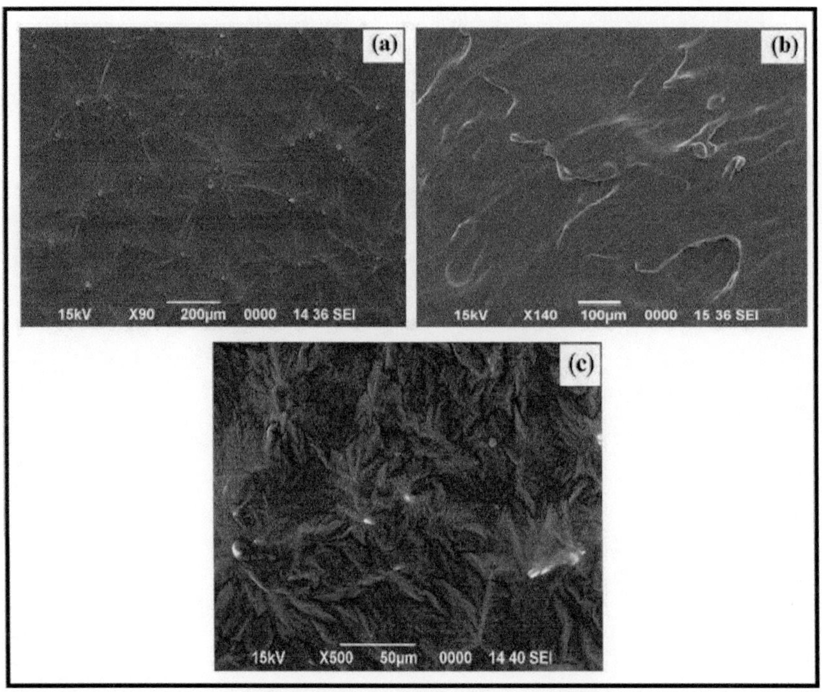

FIGURE 10.24 SEM of PHA films obtained from *P. aeruginosa* JQ796859 (a); *B. circulans* MTCC8167 (b); and *P. aeruginosa* JQ866912 (c).

KEYWORDS

- **differential scanning calorimetry**
- **photoluminescence**
- **polyhydroxyalkanoates**
- **polymers**
- **thermogravimetric**
- **total ion chromatogram**
- **X-ray diffraction**

REFERENCES

Berlanga, M., et al., (2006). Rapid spectrofluorometric screening of poly-hydroxyalkanoate-producing bacteria from microbial mats, *International Microbiology, 9*, 95–102.

Kadouri, D., et al., (2005). Ecological and agricultural significance of bacterial polyhydroxyalkanoates. *Crit. Rev. Anal. Chem., 31*, 55–67.

Williamson, D. H., & Wilkinson, J. F., (1958). The isolation and estimation of poly-β-hydroxybutyrate inclusions of *Bacillus* species. *J. Gen. Microbiol., 19*, 198–209.

CHAPTER 11

Application of PHAs

PHAs have been drawing considerable attention as biocompatible plastics for a wide range of applications. Several companies are developing PHAs for use in the plastic industry. Currently, Metabolix (USA) and Kaeka (Japan) are among the few companies have actively teamed up for the commercialization of PHA biopolymers known as Nodax. This product has already been made into a variety of different prototype objects such as plastic fiber or twine and molded plastic-ware. The extensive range of physical properties of PHA provides them a broad range of potential applications to mankind.

Yu and Chen (2006) reported the use of P(HB-HV) in making films, blow molded bottles, containers, and as a creating agent on paper. Bioplastic composites have already been used in electronic products, like mobile phones. PHA could chemically modify their functional groups as well as their biodegradability and biocompatibility properties, making them attractive material in the biomedical field (Ratledge and Kristiansen, 2001). PHAs are mainly composed of different chiral hydroxy acids that have potential as synthons for anticancer drugs, anti-HIV drugs, antibiotics, and vitamins. In biomedical and tissue engineering applications, PHAs have mainly been involved for biodegradable implants. PHA together with hydroxyapatite (HA) can find an application as a bioactive and biodegradable composite in hard tissue replacement. Kilicay et al. (2011) reported that a matrix of PHBHHx nanoparticles (NPs) has been used to deliver anti-neoplastic agents to cancer cells. Certain target specific breast cancer cells were also examined with PHB NPs functionalized with a tumor-specific ligand

Bacterial Biopolymers. B. K. Konwar, PhD (Author)

(Haywood et al., 1991). Francis (2011) tested P(3HB) microsphere for releasing the antibiotics gentamycin and tetracycline *in vitro*. The slow hydrolytic degradation inside the human body makes PHAs more advantageous for the use in the reconstructive surgery. The degradation product of PHB, D-(–)-3-hydroxybutyrate has been detected in relatively large amount in human blood plasma. PHAs can also be used as starting material for other chemicals taking advantage of their uniform chirality (Bucci and Tavares, 2005). Sudesh et al. (2000) reported that the 4HB units of PHB compounds have been used in the treatment of alcohol withdrawal syndrome. Partial digestion of PHA and recombination with other polymers can be used to achieve specific properties. 'Degra Pol,' a block-copolyester urethane chemically synthesized from PHB-diol shows good biocompatibility.

In experimental animals, P(3HBHHx) scaffolds have been assessed for use in the eyelid reconstruction (Lee et al., 2011) in tumor-specific hybrid polyhydroxybutyrate (PHB) nanoparticle: surface modification of nanoparticle by enzymatically synthesized functional block copolymer. Moreover, PHA has been used in other biomedical applications, such as tablet formulations (Zhou et al., 2010; Patel, 2011), surgical sutures (Riekes et al., 2011), wound dressings (Ra et al., 2011), and controlled release contraceptive devices (Ozturk and Ermertca, 2011; Uppal et al., 2011).

In addition to biomedical applications, PHAs have other diverse applications. A recent study suggested that PHAs can be used as precursors for biofuel production (Bansal et al., 2011). Hydrolysis of PHAs followed by methyl esterification provides energy containing 3-hydroxyalkanoates methyl esters comparable to that of bioethanol (Waknis and Jonnalagadda, 2011). Foster et al. (2001) reported that PHAs has a potential role as a pollution bio-indicator in the assessment of environmental health. It was reported by Hiraishi and Khan (2003) that many PHAs also used in the solid phase denitrification of water and wastewater has several advantages over the conventional system supplemented with liquid organic substrate. Bourbonnais and Marchessault (2010) reported the use of PHA latex in the paper industry for surface coating and as sizing agent. In certain aquaculture applications, PHAs have been potentially useful for controlling bacterial pathogens (Nhan et al., 2010). PHAs have also been used as controlled release-agent for herbicides in agriculture which can potentially reduce the repeated need of the herbicides on non-target species (De Schryver et al., 2011).

11.1 USE IN NANOCOMPOSITES

Silver nanoparticles (SNPs) have attracted much attention because of their antibacterial properties. The dominant use of SNP is in the cosmetics, antibacterial-water filters and as drug carrier. But slurries of such particles tend to be unstable and therefore always need some biodegradable but stable polymer support. In this connection, the use of a hydrophobic, biodegradable, renewable, and non-cytotoxic biopolymer for stabilizing the SNP particle would provide an advantage for the synthesis, transportation of the SNP colloids and their use in different biomedical causes. In the light of the above information, an experiment designed for stabilizing the colloidal solution of SNP by using bacterial PHAs was reported by us (Phukon et al., 2011). The hydrophobic nature of PHA would speed up the process of water filtration as the water could access the surfaces and pores of the particles where SNP would be in touch with the water. In cosmetic technology, the SNP–PHA might act as a consistent stabilizer and the stability may lead to an increase in the duration of sunscreen protection without adding excess of SNP.

11.2 USE IN BIOSENSORS

PHAs have been used in the manufacture of biosensors. Myoglobin immobilized in P(3HB) film provides a model for constructing a third generation H_2O_2 biosensor (Ma et al., 2005). The side effects from the artemisinin class of medications are like the symptoms of malaria. A case of significant liver inflammation is reported in association with the prolonged use of a relatively high dose of artemisinin for an unclear reason. In the treatment of severe malaria, parenteral artesunate has shown promising results by reducing the mortality rate among the patients in South East Asian countries by 35% as compared to quinine. Therefore, it is important and interesting to develop a simple, sensitive, fast, portable, and reliable method for the detection and quantification of artemisinin.

KEYWORDS

- **artemisinin**
- **hydrophobic nature**
- **hydroxyapatite**
- **myoglobin**
- **polyhydroxyalkanoates**
- **polymers**
- **silver nanoparticles**

REFERENCES

Bansal, S. S., et al., (2011). Development and *in vitro-in vivo* evaluation of polymeric implants for continuous systemic delivery of curcumin. *Pharm. Res., 28*, 1121–1130.

Bourbonnais, R., & Marchessault, R. H., (2010). Application of polyhydroxyalkanoates granules for sizing of paper. *Biomacromolecules, 11*, 989–993.

Bucci, D. Z., & Tavares, L. B. B., (2005). PHB packaging for the storage of food products. *Polym. Test., 24*, 564–571.

De Schryver, P., et al., (2011). Convergent dynamics of the juvenile European sea bass gut microbiota induced by polyhydroxybutyrate. *Environ. Microbiol., 13*, 1042–1051.

Defoirdt, T., et al., (2011). Alternatives to antibiotics for the control of bacterial disease in aquaculture. *Curr. Opin. Microbiol., 14*, 251–258.

Flieger, M., et al., (2003). Biodegradable plastics from renewable sources. *Folia Microbiol., 48*, 2–44.

Foster, L. J. R., et al., (2001). Environmental concentrations of polyhydroxyalkanoates and their potential as bioindicators of pollution. *Biotechnol. Lett., 23*, 893–898.

Francis, L., (2011). *Biosynthesis of Polyhydroxyalkanoates and Their Medical Applications.* Ph. D. Thesis, University of Westminster, Westminster.

Grillo, R., et al., (2011). Controlled release system for ametryn using polymer microspheres: Preparation, characterization and release kinetics in water. *J. Hazard. Mater., 186*, 1645–1651.

Gumel, A. M., et al., (2012). Recent advances in the production, recovery and applications of polyhydroxyalkanoates. *J. Polym. Environ., 21*, 580–605.

Hiraishi, A., & Khan, S. T., (2003). Application of polyhydroxyalkanoates for denitrification in water and wastewater treatment. *Appl. Microbiol. Biotechnol., 61*,103–109.

Kilicay, E., et al., (2011). Preparation and characterization of poly (3-hydroxybutyrate-co-3-hydroxyhexanoate) (Phbhhx) based nanoparticles for targeted cancer therapy. *Eur. J. Pharm. Sci., 44*, 310–320.

Lee, J., et al., (2011). Tumor-specific hybrid polyhydroxybutyrate nanoparticle: Surface modification of nanoparticle by enzymatically synthesized functional block copolymer. *Bioorg. Med. Chem. Lett., 21*, 2941–2944.

Ma, X., et al., (2005). Hydrogen peroxide biosensor based on the direct electrochemistry of myoglobin immobilized in poly-3-hydroxybutyrate film. *American J. of Biochemistry and Biotechnology, 1*, 43–46.

Nhan, D. T., et al., (2010). The effect of poly [beta]-hydroxybutyrate on larvi culture of the giant freshwater prawn *Macrobrachium rosenbergii*. *Aquaculture, 302*, 76–81.

Ozturk, F., & Ermertcan, A. T., (2011). Wound healing: A new approach to the topical wound care. *Cutan. Ocul. Toxicol., 30*, 92–99.

Patel, A. R., & Velikov, K. P., (2011). Colloidal delivery systems in foods: A general comparison with oral drug delivery. *LWT- Food Sci. Technol., 44*, 1958–1964.

Phukon, P., et al., (2011). Enhancing the stability of colloidal silver nanoparticles using polyhydroxyalkanoates (PHA) from *Bacillus circulans* (MTCC 8167) isolated from crude oil contaminated soil. *Colloids Surf., B, 86*, 314–318.

Ra, R., et al., (2011). Medium chain length polyhydroxyalkanoates, promising new biomedical materials for the future, *Mater. Sci. Eng. R: Reports, 72*, 2947.

Riekes, M. K., et al., (2011). Evaluation of oral carvedilol microparticles prepared by simple emulsion technique using poly (3-hydroxybutyrate-co-3-hydroxyvalerate) and polycaprolactone as polymers. *Mater. Sci. Eng. C, 31*, 962–968.

Sudesh, K., et al., (2000). Synthesis, structure and properties of polyhydroxyalkanoates: Biological polyesters. *Prog. Polym. Sci., 25*, 1503–1555.

Uppal, R., et al., (2011). Hyaluronic acid nanofiber wound dressing—production, characterization, and *in vivo* behavior. *J. Biomed. Mater. Res. B Appl. Biomater., 97*, 20–29.

Waknis, V., & Jonnalagadda, S., (2012). Novel poly-DL-lactidepolycaprolactone copolymer based flexible drug delivery system for sustained release of ciprofloxacin. *Drug Delivery, 18*, 236–245.

Zhou, J., et al., (2010). The use of poly (3-hydroxybutyrate-co-3-hydroxyhexanoate) scaffolds for tarsal repair in eyelid reconstruction in the rat. *Biomaterials, 31*, 7512–7518.

CHAPTER 12

Biopolymer-Producing Bacteria and the Genes

12.1 IDENTIFICATION OF PHA BIOSYNTHETIC GENES

The genetic organization of PHA biosynthetic genes varies among PHA-producing organisms. The type II of PHA biosynthetic genetic system consists of two PhaC1 and PhaC2 genes, separated by the enzyme depolymerase phaZ gene. The type II system is commonly found in Mcl-PHA-producing pseudomonads. A rapid and sensitive PCR procedure was used for the detection of PhaC genes using primers, I-179L and I-179R, on the basis of conserved sequences in the coding regions of PhaC1 and PhaC2 genes present in *Pseudomonas* (1997). The PHA biosynthetic gene locus, PhaC1 in *Pseudomonas resinovorans* and the same was subsequently PCR-based cloning and expressed in *E. coli* as reported by Solaiman et al. (2000). Zhang et al. (2001) designed a PCR-based cloning approach using highly conserved regions of PHA biosynthesis gene locus to clone the PHA biosynthetic genes from Pseudomonads. Solaiman (2002 a, b) used a semi-nested PCR method for the specific amplification of type II sub-genomic fragments PhaC1 and PhaC2. The method was used to show that the strains of *Pseudomonas oleovorans* harbor different PHA loci. Jamil et al. (2007) reported PCR-based amplification of PHA polymerase genes C_1 and C_2 from chromosomal DNA and apportion of these genes in the PHA operon was cloned and sequenced.

Bacterial Biopolymers. B. K. Konwar, PhD (Author)

12.2 ISOLATION OF MICROORGANISMS BY ENRICHMENT CULTURE

The bacterial strains derived from the crude petroleum contaminated soil and water; petroleum sludge samples were cultured using enrichment culture technique. A 5 g sample in 100 ml of minimum salt medium (MSM) was cultured according to Bordoloi and Konwar (2008) in 500 ml Erlenmeyer flask supplemented with 1% (v/v) of membrane filtered (0.22 m) n-hexadecane as the carbon source and incubated at 37°C in orbital incubator shaker for 7 days at 180 rpm. After 10 cycles of enrichment, 1 ml of the saturated culture was diluted 104 times and an aliquot of 100 L was spread on MSM agar plates having 0.1% (v/v) n-hexadecane. The culture plates were incubated at 37°C for 48 h; morphologically distinct bacterial colonies thus obtained were further purified on the MSM-agar plates with or without 0.1% (v/v) n-hexadecane to eliminate autotrophs and agar-utilizing bacteria. The procedure was repeated and isolates showing pronounced growth on n-hexadecane were selected and preserved for further characterization (Bordoloi and Konwar, 2008).

12.3 ISOLATION OF MICROORGANISMS BY DIRECT CULTURE TECHNIQUE

The crude oil contaminated soil, water, and petroleum sludge samples were serially diluted with sterile saline water (0.9% NaCl, w/v) using the standard dilution technique. The total viable bacterial populations were determined by spread-plating each sample after appropriate dilution (105,106-fold) in nutrient agar. The morphologically different bacterial colonies thus obtained were further purified on nutrient agar. Bacterial isolates were further grown on Bushnell-Hass medium (2008) to ensure their hydrocarbon utilizing capacity. The medium was supplemented with 1% (v/v) n-hexadecane and was used for the screening of potential hydrocarbon-degrading and biosurfactant producing bacterial isolates (Bordoloi and Konwar, 2008).

The bacterial pure cultures were preserved at 4°C in both nutrient agar plates and slants. Cultures were sub-cultured at an interval of 30 days. For long time storage, frozen stock cultures were prepared in 15% (v/v) glycerol and stored at 70°C.

12.4 PHA PRODUCING BACTERIA IN HYDROCARBON-BASED ENVIRONMENTAL SAMPLES

The environmental samples consist of crude oil-contaminated soil of oil fields, petroleum sludge and waste residual crude oil dumping sites from ONGC, Jorhat, Assam, India; petroleum sludge from the oil fields of ONGC, Sibsagar; contaminated soil samples from the different oil depots of Tezpur, Assam. The bacterial strains were isolated from the sated environmental samples using both enrichment and direct plate culture techniques. A total of 52 culturable isolates were obtained based on distinctly different colony morphology as shown in Table 12.1.

TABLE 12.1 Morphological Characters of Bacterial Isolates Obtained from Crude Oil-Contaminated Samples

SL. No.	Bacterial Isolates	Size	Pigment	Form	Margin	Elevation
1.	JB08S11	Large	Green	Irregular	Lobate	Flat
2.	JB08S12	Large	White	Circular	Entire	Flat
3.	JB08S13	Large	Yellow	Circular	Undulate	Flat
4.	JB08S14	Moderate	White	Circular	Entire	Raised
5.	JB08S15	Moderate	White	Irregular	Lobate	Raised
6.	JB08S16	Small	White	Circular	Entire	Raised
7.	JB08S17	Small	Light yellow	Circular	Entire	Raised
8.	JB08S18	Small	White	Circular	Entire	Flat
9.	JB08S21	Large	Light yellow	Irregular	Lobate	Flat
10.	JB08S22	Large	White	Circular	Undulate	Flat
11.	JB08S23	Small	White	Circular	Entire	Raised
12.	JB08S24	Small	White	Irregular	Lobate	Flat
13.	JB08S25	Pinpoint	White	Circular	Entire	Raised
14.	JB08S31	Large	Yellow	Circular	Entire	Convex
15.	JB08S32	Large	White	Circular	Entire	Raised
16.	JB08S33	Large	Green	Circular	Undulate	Flat
17.	JB08S34	Large	Yellow	Irregular	Lobate	Flat
18.	JB08PS35	Moderate	White	Irregular	Lobate	Raised
19.	JB08PS36	Moderate	White	Irregular	Lobate	Raised
20.	JB08PS37	Moderate	White	Circular	Entire	Raised

TABLE 12.1 *(Continued)*

SL. No.	Bacterial Isolates	Size	Pigment	Form	Margin	Elevation
21.	JB08PS38	Moderate	Green	Circular	Serrate	Raised
22.	JB08PS39	Pinpoint	White	Circular	Entire	Flat
23.	JB08PS41	Moderate	White	Circular	Undulate	Flat
24.	JB08PS42	Pinpoint	White	Circular	Entire	Raised
25.	JB10OD11	Large	Light yellow	Irregular	Lobate	Flat
26.	JB10OD12	Large	White	Rhizoid	Filamentous	Flat
27.	JB10OD13	Moderate	Yellow	Irregular	Lobate	Raised
28.	JB10OD14	Moderate	White	Irregular	Entire	Raised
29.	JB10OD15	Small	White	Circular	Serrate	Convex
30.	JB10OD21	Large	White	Circular	Entire	Flat
31.	JB10OD22	Large	Yellow	Irregular	Lobate	Raised
32.	JB10OD23	Moderate	White	Circular	Entire	Raised
33.	JB10OD24	Moderate	Yellow	Circular	Entire	Convex
34.	JB10OD25	Pinpoint	White	Circular	Lobate	Raised
35.	S10PSS11	Large	Light yellow	Rhizoid	Filamentous	Flat
36.	S10PSS12	Large	White	Circular	Entire	Convex
37.	S10PSS13	Moderate	White	Circular	Entire	Convex
38.	S10PSS14	Moderate	Yellow	Circular	Undulate	Flat
39.	S10PSS15	Moderate	White	Circular	Entire	Raised
40.	S10PSS16	Small	White	Circular	Serrate	Flat
41.	S10PSS17	Small	Yellow	Circular	Entire	Raised
42.	S10PSS21	Large	White	Circular	Serrate	Raised
43.	S10PSS22	Large	White	Irregular	Lobate	Flat
44.	S10PSS23	Moderate	White	Circular	Entire	Convex
45.	S10PSS24	Small	Green	Circular	Entire	Flat
46.	S10PSS25	Pinpoint	Green	Circular	Entire	Flat
47.	T11PS11	Moderate	White	Circular	Entire	Flat
48.	T11PS12	Moderate	White	Circular	Serrate	Flat
49.	T11PS21	Large	Green	Circular	Entire	Flat
50.	T11PS22	Small	Yellow	Circular	Entire	Raised
51.	T11PS23	Small	Yellow	Circular	Entire	Raised
52.	T11PS24	Pinpoint	White	Circular	Entire	Raised

The bacterial isolates were further cultured to obtain the pure colonies of each individual strain and maintained in nutrient plates, also in stab agar cultures at 4°C. The isolates were sub-cultured at an interval of 30 days in nutrient broth (NB) agar. For long term maintenance, the bacterial isolates were preserved in 15% (w/v) glycerol and stored at 70°C.

After the initial screening at 37°C, all 52 bacterial strains were re-cultured in 100 ml mineral salt medium supplemented with 1% (v/v) of n-hexadecane as the sole source of carbon and incubated at 37°C for 7 days at a constant rpm of 150. After the sixth cycle of subcultures, the bacterial isolates were cultured on Bushnell-Hass medium (2008) supplemented with 1% (v/v) n-hexadecane to establish their hydrocarbon utilizing ability. The bacterial isolates were grown for 96 h at 37°C with 150 rpm to determine the total dry biomass yield and the same are presented in Table 12.2.

TABLE 12.2 Biomass Yields in Bushnell-Hass Medium Supplemented with 1% (v/v) n-Hexadecane after 96 h

SL. No.	Bacterial Isolates	Biomass ($g.l^{-1}$)
1.	JB08S11	2.32 ± 0.75
2.	JB08S12	0.45 ± 0.32
3.	JB08S13	1.87 ± 0.43
4.	JB08S14	4.78 ± 0.21
5.	JB08S15	4.23 ± 0.65
6.	JB08S16	2.15 ± 0.31
7.	JB08S17	4.54 ± 0.65
8.	JB08S18	0.20 ± 0.55
9.	JB08S21	4.93 ± 0.83
10.	JB08S22	2.8 ± 0.32
11.	JB08S23	4.46 ± 0.27
12.	JB08S24	2.83 ± 0.11
13.	JB08S25	5.02 ± 0.45
14.	JB08S31	4.93 ± 0.83
15.	JB08S32	2.20 ± 0.29
16.	JB08S33	4.75 ± 0.43
17.	JB08S34	0.32 ± 1.2
18.	JB08PS35	5.10 ± 0.82
19.	JB08PS36	0.71 ± 0.37

TABLE 12.2 *(Continued)*

SL. No.	Bacterial Isolates	Biomass (g.l^{-1})
20.	JB08PS37	1.65 ± 0.44
21.	JB08PS38	4.77 ± 0.62
22.	JB08PS39	2.53 ± 0.83
23.	JB08PS41	0.16 ± 0.97
24.	JB08PS42	4.92 ± 1.1
25.	JB10OD11	5.08 ± 0.41
26.	JB10OD12	0.67 ± 0.67
27.	JB10OD13	2.83 ± 0.93
28.	JB10OD14	4.86 ± 0.83
29.	JB10OD15	4.25 ± 0.53
30.	JB10OD21	3.52 ± 0.77
31.	JB10OD22	5.02 ± 0.84
32.	JB10OD23	3.94 ± 0.53
33.	JB10OD24	2.17 ± 0.91
34.	JB10OD25	4.94 ± 0.63
35.	S10PSS11	5.05 ± 0.82
36.	S10PSS12	3.32 ± 0.32
37.	S10PSS13	4.93 ± 0.54
38.	S10PSS14	4.05 ± 0.82
39.	S10PSS15	4.88 ± 0.43
40.	S10PSS16	3.62 ± 0.81
41.	S10PSS17	5.09 ± 0.80
42.	S10PSS21	3.58 ± 0.31
43.	S10PSS22	4.95 ± 0.52
44.	S10PSS23	0.07 ± 1.1
45.	S10PSS24	0.27 ± 0.62
46.	S10PSS25	5.02 ± 0.41
47.	T11PS11	4.75 ± 0.53
48.	T11PS12	1.88 ± 0.41
49.	T11PS21	3.91 ± 0.74
50.	T11PS22	5.10 ± 0.18
51.	T11PS23	4.90 ± 0.42
52.	T11PS24	2.96 ± 0.62

Note: Results represent mean ± S.D of three individual experiments.

A total of 23 potential isolates JB08S14, JB08S17, JB08S21, JB08S23, JB08S25, JB08S31, JB08S33, JB08PS35, JB08PS38, JB08PS42, JB10OD11, JB10OD14, JB10OD22, JB10OD25, S10PSS11, S10PSS13, S10PSS15, S10PSS17, S10PSS22, S10PSS25, T11PS11, T11PS22, and T11PS23 possessing better growth in Bushnell-Hass medium supplemented with 1% (v/v) n-hexadecane were marked based on increased dry biomass yield. The isolates were also screened for the production of biosurfactant by culturing on n-hexadecane supplemented medium. The colony morphology of the bacterial isolates is shown in Figure 12.1. Four bacterial isolates belonging to *Pseudomonas* were revealed through characterization details revealed.

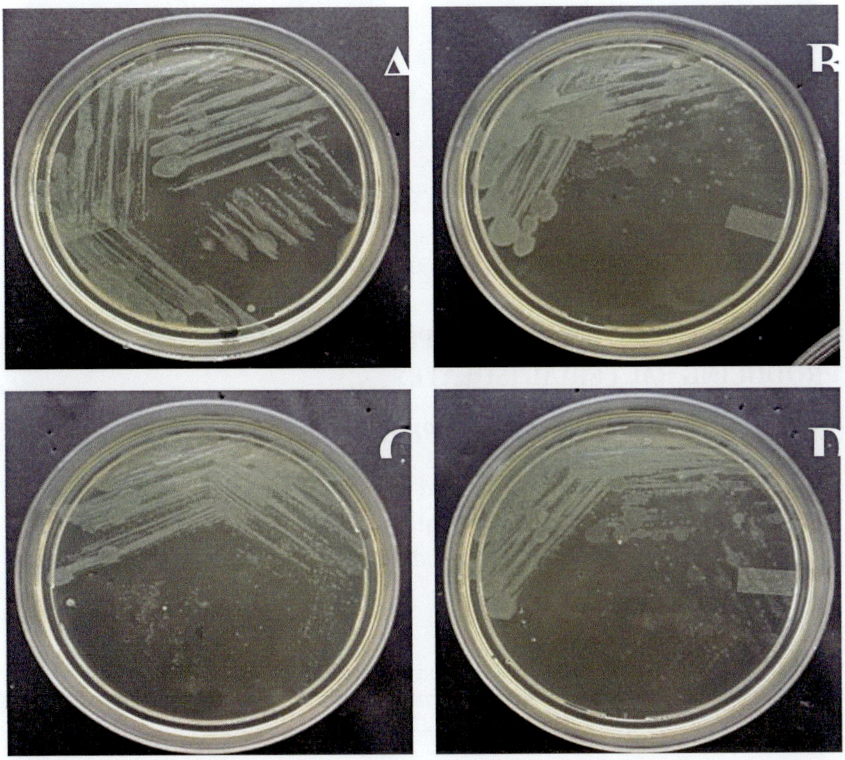

FIGURE 12.1 Colony morphology of pure cultures on nutrient agar: (A) *P. aeruginosa* OBP1; (B) *P. aeruginosa* OBP2; (C) *P. aeruginosa* OBP3; and (D) *P. aeruginosa* OBP4.

The SEM of *P. aeruginosa* strains OBP1, OBP2, OBP3 and OBP4 showing growth on n-hexadecane is presented in Figure 12.2.

FIGURE 12.2 Scanning electron micrograph (SEM) of *P. aeruginosa* strains showing growth on n-hexadecane: (a) *P. aeruginosa* OBP1; (b) OBP2; (c) OBP3; and (d) OBP4.

12.5 CHARACTERIZATION OF POTENTIAL BIOSURFACTANT PRODUCING BACTERIAL ISOLATES

12.5.1 MORPHOLOGICAL AND PHYSIOLOGICAL CHARACTERS

The biochemical, morphological, and physiological characterization of the bacterial isolates was carried out following the procedures described by Cappuccino and Sherman (1999) and data thus obtained are presented in Table 12.3.

TABLE 12.3 Biochemical Characterization of Bacterial Isolates

Biochemical Test	Bacterial Isolate			
	OBP1	OBP2	OBP3	OBP4
1. Gram staining	Negative	Negative	Negative	Negative
2. Shape of the cell	Straight rod	Straight rod	Straight rod	Straight rod
3. Capsule staining	Positive	Positive	Positive	Positive
4. Endospore staining	Negative	Negative	Negative	Negative
5. Motility test	Positive	Positive	Positive	Positive

TABLE 12.3 *(Continued)*

Biochemical Test	Bacterial Isolate			
	OBP1	OBP2	OBP3	OBP4
6. Acid production from:				
a. Glucose	Positive	Positive	Positive	Positive
b. Fructose	Positive	Positive	Positive	Positive
c. Xylose	Positive	Positive	Positive	Positive
d. Maltose	Negative	Negative	Negative	Negative
e. Lactose	Negative	Negative	Negative	Negative
f. Mannitol	Negative	Negative	Negative	Negative
g. Salicin	Negative	Negative	Negative	Negative
h. Sucrose	Negative	Negative	Negative	Negative
i. Oxidase test	Positive	Positive	Positive	Positive
7. Catalase test	Positive	Positive	Positive	Positive
8. H_2S production	Negative	Negative	Negative	Negative
9. Gelatin hydrolysis test	Positive	Positive	Positive	Positive
10. Starch hydrolysis	Negative	Negative	Negative	Negative
11. Methyl red test	Negative	Negative	Negative	Negative
12. Voges-Proskauer test	Negative	Negative	Negative	Negative
13. Indole test	Negative	Negative	Negative	Negative
14. Citrate utilization test	Positive	Positive	Positive	Positive
15. Nitrate reduction test	Positive	Positive	Positive	Positive
16. Production of fluorescence	Positive	Positive	Positive	Positive
17. Production of pyocyanin	Negative	Negative	Negative	Negative

12.5.2 MOLECULAR CHARACTERIZATION OF BACTERIAL ISOLATES

The selected bacterial isolates were subjected to partial sequencing of their 16S-rRNA genes, the partial gene sequences as well as with the use of NCBI GenBank BLAST tool, these four bacterial isolates OBP1, OBP2, OBP3, and OBP4 were close to *Pseudomonas aeruginosa* with 99% similarity. The BLAST analysis on the NCBI database on the basis of score, query coverage, E-value, and maximum identity showed the maximum similarity with 34 different strains of *P. aeruginosa*. The partial 16S-rRNA gene sequences of OBP1, OBP2, OBP3, and OBP4 were deposited in the GenBank with the accession numbers 1568190, 1568199, 1568206, and 156820, respectively. Further, putative phylogenetic trees (Saitou and Nei, 1987) for the bacterial isolates were constructed using the Neighbor-Joining method and are shown in Figure 12.3.

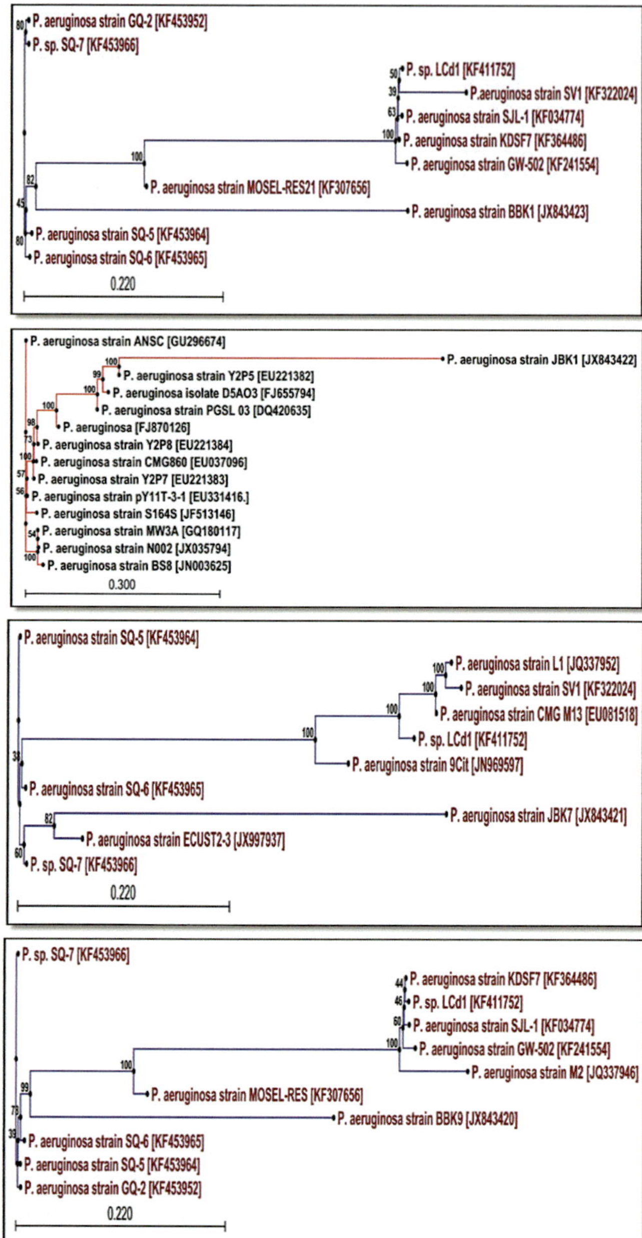

FIGURE 12.3 Phylogenetic tree generated using neighbor-joining method showing the similarity of selected strains with other 16S-rRNA gene sequences of *P. aeruginosa*. Bootstrap values are expressed as percentages of 1,000 replications. Bar, 0.01 substitutions per nucleotide position. *P. aeruginosa* (A) OBP1; (B) OBP2; (C) OBP3; and (D) OBP4.

The CLC main workbench was used in the nucleotide frequency count was done for these isolates. The nucleotide frequency counts showed that ATGC, C+G and A+T compositions of bacterial strains OBP1, OBP2, OBP3, and OBP4 have almost similar frequency (99%) distribution as compared to other reported strains of *P. aeruginosa*.

12.6 ABILITY OF THE BACTERIAL STRAINS TO GROW IN DIFFERENT HYDROCARBONS

The bacterial strains were able to grow in MSM supplemented with different hydrocarbons as the sole source of carbon and energy. The growth performance of the bacterial strains on the hydrocarbon sources has been presented in Table 12.4. The bacterial strains exhibited better growth on n-hexadecane, octadecane, tridecane, dodecane, diesel, and crude oil supplemented media. Media containing kerosene, lubricating oil, paraffin, pentane, hexane, heptane, iso-octane, eicosane, and triacontane possessed little or no growth.

TABLE 12.4 Ability of the Bacterial Strains to Utilize Different Components of Crude Petroleum

Carbon Sources	Properties	*P. aeruginosa* Strains			
		OBP1	OBP2	OBP3	OBP4
Pentane	DB $g.l^{-1}$	–	–	–	–
	ST mNm^{-1}	68.3 ± 0.21	67.5 ± 0.42	68.8 ± 0.43	67.9 ± 0.32
Hexane	DB $g.l^{-1}$	–	–	–	–
	ST mNm^{-1}	67.9 ± 0.32	68.5 ± 0.61	68.1 ± 0.45	67.7 ± 0.32
Heptane	DB $g.l^{-1}$	–	–	–	–
	ST mNm^{-1}	68.4 ± 0.24	67.8 ± 0.57	67.6 ± 0.31	68.3 ± 0.36
Iso-octane	DB $g.l^{-1}$	–	–	–	–
	ST mNm^{-1}	67.9 ± 0.36	68.2 ± 0.51	67.7 ± 0.48	67.9 ± 0.52
Dodecane	DB $g.l^{-1}$	3.28 ± 0.65	3.31 ± 0.34	2.55 ± 0.53	2.37 ± 0.65
	ST mNm^{-1}	36.8 ± 0.21	43.4 ± 0.63	41.6 ± 0.27	38.9 ± 0.52
Tridecane	DB $g.l^{-1}$	3.37 ± 0.64	3.48 ± 0.45	3.61 ± 0.44	3.54 ± 0.32
	ST mNm^{-1}	36.0 ± 0.24	42.5 ± 0.57	39.7 ± 0.31	37.5 ± 0.36
n-Hexadecane	DB $g.l^{-1}$	4.87 ± 0.63	5.03 ± 0.37	4.73 ± 0.72	5.10 ± 0.21
	ST mNm^{-1}	31.1 ± 0.88	37.6 ± 0.51	35.5 ± 0.38	33.2 ± 0.79

TABLE 12.4 *(Continued)*

Carbon Sources	Properties	*P. aeruginosa* Strains			
		OBP1	OBP2	OBP3	OBP4
Octadecane	DB $g.l^{-1}$	4.83 ± 0.34	3.82 ± 0.56	4.12 ± 0.74	4.24 ± 0.37
	ST mNm^{-1}	31.9 ± 0.85	39.6 ± 0.24	38.4 ± 0.42	36.8 ± 0.65
Eicosane	DB $g.l^{-1}$	–	–	–	–
	ST mNm^{-1}	68.6 ± 0.18	68.4 ± 0.61	67.9 ± 0.28	68.2 ± 0.20
Triacontane	DB $g.l^{-1}$	–	–	–	–
	ST mNm^{-1}	67.7 ± 0.49	68.3 ± 0.61	68.5 ± 0.28	67.9 ± 0.19
Paraffin	DB $g.l^{-1}$	1.38 ± 0.22	1.13 ± 0.46	1.52 ± 0.27	1.47 ± 0.51
	ST mNm^{-1}	45.3 ± 0.29	47.2 ± 0.27	44.5 ± 0.25	45.6 ± 0.55
Phenol	DB $g.l^{-1}$	–	–	–	–
	ST mNm^{-1}	68.3 ± 0.23	68.6 ± 0.23	67.8 ± 0.27	67.4 ± 0.82
Benzene	DB $g.l^{-1}$	–	–	–	–
	ST mNm^{-1}	68.3 ± 0.18	68.6 ± 0.23	67.9 ± 0.41	68.4 ± 0.21
Toluene	DB $g.l^{-1}$	0.36 ± 0.13	–	0.78 ± 0.52	1.04 ± 0.22
	ST mNm^{-1}	54.7 ± 0.39	68.7 ± 0.41	52.2 ± 0.17	52.5 ± 0.48
Xylene	DB $g.l^{-1}$	–	–	–	–
	ST mNm^{-1}	68.2 ± 0.16	68.5 ± 0.42	68.7 ± 0.19	68.5 ± 0.15
Naphthalene	DB $g.l^{-1}$	–	–	0.83 ± 0.66	–
	ST mNm^{-1}	68.7 ± 0.69	68.2 ± 0.36	53.5 ± 0.35	68.7 ± 0.92
Anthracene	DB $g.l^{-1}$	0.45 ± 0.37	–	0.66 ± 0.27	0.93 ± 0.41
	ST mNm^{-1}	54.9 ± 0.83	67.7 ± 0.23	54.2 ± 0.44	53.6 ± 0.58
Phenanthrene	DB $g.l^{-1}$	0.61 ± 0.34	–	1.05 ± 0.29	1.18 ± 0.73
	ST mNm^{-1}	54.2 ± 0.59	68.6 ± 0.15	53.8 ± 0.43	53.2 ± 0.46
Pyrene	DB $g.l^{-1}$	–	–	–	–
	ST mNm^{-1}	67.8 ± 0.91	68.7 ± 0.20	67.6 ± 0.82	68.7 ± 0.73
Fluorene	DB $g.l^{-1}$	–	–	–	–
	ST mNm^{-1}	68.8 ± 0.83	67.9 ± 0.93	68.6 ± 0.43	67.9 ± 0.42
Diesel	DB $g.l^{-1}$	4.54 ± 0.93	3.97 ± 0.53	4.91 ± 0.40	5.04 ± 0.62
	ST mNm^{-1}	32.0 ± 0.42	37.5 ± 0.28	36.2 ± 0.69	34.3 ± 1.0
Kerosene	DB $g.l^{-1}$	2.73 ± 0.76	2.30 ± 0.56	2.54 ± 3.2	2.78 ± 0.74
	ST mNm^{-1}	39.6 ± 0.62	42.2 ± 0.29	40.5 ± 0.83	39.2 ± 0.44
Lubricating oil	DB $g.l^{-1}$	2.31 ± 0.56	1.67 ± 0.21	2.39 ± 0.82	2.24 ± 0.29
	ST mNm^{-1}	41.5 ± 0.37	43.8 ± 0.18	40.7 ± 0.64	41.4 ± 1.0
Crude oil	DB $g.l^{-1}$	3.71 ± 0.44	3.27 ± 0.62	3.86 ± 0.82	4.07 ± 0.53
	ST mNm^{-1}	32.7 ± 0.66	40.4 ± 0.25	39.5 ± 0.41	37.7 ± 0.16

Note: Results represent mean ± S.D of three individual experiments.

In the case of aromatic hydrocarbons like toluene, the bacterial strains OBP4, OBP3, and OBP1 showed slight growth but no growth on benzene, phenol, and xylene supplemented media. The strains OBP4, OBP3, and OBP1 exhibited minimum growth on PAHs like phenanthrene and anthracene but no growth on pyrene and fluorene. Out of the four bacterial strains, only OBP3 exhibited growth on naphthalene supplemented medium. The n-hexadecane supplemented MSM exhibited good growth. The capability of the bacterial strains to utilize a hydrocarbon as the sole source of carbon and energy was different from each other as revealed by their biomass yield and efficiency to reduce the surface tension.

12.7 MICROBIAL STRAINS PRODUCING BIOPOLYMERS

The biopolymer producing bacterium *Bacillus circulan*s MTCC8167 was previously isolated in our laboratory from crude oil contaminated soil and taxonomically identified. Different media were used during the investigation, and the same are listed below. The maintenance medium is a liquid or solid designed to maintain the growth of microorganisms. The medium contains all essential nutrient elements for almost all bacterial growth, non-selective, and is used for culture in general as well as maintenance and then kept in laboratory culture collections.

- Medium used: Nutrient agar (pH 7);
- Peptone: 5.0 $g.l^{-1}$;
- NaCl: 5.0 $g.l^{-1}$;
- Beef extract: 3.0 $g.l^{-1}$;
- Agar: 15 $g.l^{-1}$ of distilled water.

The inoculum medium provides an appropriate biochemical and biophysical environment for the cultivation and propagation of bacteria. The medium is also used in the isolation and maintenance of bacterial pure cultures.

- Medium: Luria Bertani medium (pH 7.0);
- Tryptone: 10.0 gL^{-1};
- NaCl: 10.0 gL^{-1}; and
- Yeast extract: 5.0 gL^{-1} of distilled water.

12.7.1 PRODUCTION MEDIUM

The production medium is used for the growth of some bacteria containing specific substances which can allow the growth of the desired species. This type of media is also useful for identifying unknown bacteria. PHA detection medium (pH 7.0) (Modified by Rehman et al., 2007): $(NH_4)2SO_4$: 1.0 $g.l^{-1}$; KH_2PO_4: 13.3 g; $MgSO_4$: 1.3 g; Citric acid: 1.7 g; Dextrose: 10 g.l; Trace element: 10 ml [$FeSO_4 \cdot 7H_2O$ (10); $ZnSO_4 \cdot 7H_2O$ (2.25); $CuSO_4 \cdot 5H_2O$ (1); $MnSO_4 \cdot 5H_2O$ (0.5); $CaCl_2 \cdot 2H_2O$ (2.0); $Na_2B_4O_7 \cdot 10H_2O$ (0.23); $(NH_4)_6MO_7O_{24}$ (0.1); HCl (0.1 N) $g.l^{-1}$] of distilled water.

Media used for various experiments were sterilized in flasks/tubes plugged with non-absorbent cotton, at 121°C and 15 psi steam pressure for 20 min. Heat sensitive things were sterilized by proper filtration. Glasswares were sealed with autoclavable polypropylene bags before sterilization.

12.7.2 ISOLATION OF BIOPOLYMER PRODUCING BACTERIA AND THEIR CULTURE

The bacterial isolates were isolated from the crude oil contaminated soils of ONGC Assam and Assam-Arkan Basin (Jorhat); NRL waste disposal tepit (i.e., waste disposal tank) (Golaghat); and Jagirod Paper Mill (Nagaon), Assam. Soil samples were collected at random in sterile 50 ml polystyrene tubes. A minimum of five samples were collected from each location. A clean shovel was used to take the sample from an average depth of 15 cm below the surface. This was done to avoid surface bacteria that were likely to be totally aerobic. Each sample was labeled properly with all the related information.

Pure culture of the isolated bacterial strains was established by following the standard procedure. At first, a loopful of bacterial culture was inoculated in NB with pH adjusted to 7 with a 0.5 growth OD at 600 nm. After that 100 μl of culture from the above culture was mixed in 0.9% (w/v) sterile normal saline and serially diluted up to 10^{-7}. From the above dilution, 100 μl aliquot was taken and spread over sterile nutrient agar plates and kept for 24 h at 37°C to obtain single distinct colonies.

For the isolation of single pure colony both Spread-plate and Streak-plate methods were used.

The pure cultures were preserved at 4°C in nutrient agar slants for short time (3–6 months) storage and subcultured by transferring those to fresh slants at an interval of one month. The isolates were also stored in 15% (v/v) glycerol in NB and kept at–80°C for long term storage.

12.8 TAXONOMIC IDENTIFICATION OF BIOPOLYMER PRODUCING BACTERIA

The selected biopolymer producing bacteria were taxonomically identified by: (i) standard biochemical tests; (ii) studying their morphological characteristics; and (iii) ribotyping.

Inoculum of the isolated bacterial strains was prepared by transferring a single colony from a 24 h old culture plate into 5 ml of NB or LB broth and allowed to incubate at 180 rpm and 37°C overnight. This seed culture was used for inoculating the production medium at 105 v/v. Several viable cell colonies were presented as CFU/mL by counting with a hemocytometer.

12.9 POLYHYDROXYALKANOATE (PHA) PRODUCTION

The pre-culture inoculum 10 ml was transferred to 100 mL sterile PHA medium with the glucose carbon source, kept at 37°C and 180 rpm for 48 h. Bacterial growth in the medium was indicated by the increase in the turbidity of the culture medium, the cultures were harvested and assayed for PHA production.

12.9.1 GRAM'S STAINING FOR BIOPOLYMER

i. **Crystal Violet:** Crystal violet 2 g was dissolved in 20 ml 95% ethanol and the volume was made up by distilled water to 100 mL.
ii. **Gram's Iodine:** Iodine 1 g was dissolved in 300 ml of potassium iodide solution (2 gm/300 ml).
iii. **Decolorizer:** Absolute alcohol (95% ethyl alcohol).

iv. **Safranin Stain:** Safranine 2 g was dissolved in 20 ml 95% ethyl alcohol and volume was made up by distilled water to 100 ml.

The process:

A thin smear of the 24 h bacterial culture was prepared on a clean glass slide and kept for 5–10 min for air drying followed by heat fixing. Few drops of crystal violet were flooded over the bacterial culture and allowed to stand for 1 min. The excess stain was washed off with distilled water. Gram's iodine was poured over it after primary staining and kept for 1 min. The smear was rinsed under tap water followed by addition of few drops of decolorizer over the bacterial culture to remove the excess stain. It was washed again to remove the decolorizer and followed by addition of counter stain safranine for 45 seconds. The counterstain was removed with tap water and the slide was kept at room temperature for air-drying. The bacteria were observed under 100X of a compound microscope.

12.9.2 INTRACELLULAR PHA DETECTION BY STAINING WITH SUDAN BLACK B

The bacterial colonies grown on nutrient agar plates were transferred by replica plating to nitrogen-deficient medium but containing carbon and incubated at 37°C for 3 days. Sudan Black B solution (0.02% w/v in 96% ethanol) was prepared and stained for 20 min. The dye was removed after 20 min and the plates were treated for 1 min with 10 ml of 96% ethanol. Colonies containing PHA rich cells retained the dye and appeared dark blue.

12.9.3 STAINING OF BACTERIAL CELLS WITH NILE BLUE A

The presence of PHA granules inside the bacterial cells was confirmed by Nile blue A-stained fluorescence microscopy following the method of Ostle and Holt (1982). A loopful of the bacterial cells was heat-fixed and stained with 1% aqueous stain solution. After staining, the slides were washed with tap water followed by 1 min 8% acetic acid, rewashed, and the smear was blotted dry and covered with glass cover slip. Fluorescence microscopy at excitation WL 460 nm was done to detect the presence of PHA granules.

12.9.4 MORPHOLOGICAL CHARACTERIZATION OF BACTERIA

The bacterial isolates were characterized by observing their different morphological traits, such as: size and shapes, forms, margins, elevation, and different pigments.

12.9.5 BIOCHEMICAL CHARACTERIZATION OF BACTERIA

Biochemical tests conducted for the biochemical characterization of the bacterial isolates were described as in subsections.

12.9.5.1 CATALASE TEST

The test determines the presence of catalase enzyme in the bacterial isolates. Trypticase soy agar (TSA) slant 10 ml was streaked with a loopful of 24 h bacterial culture and kept at 37°C for 48 h. Hydrogen peroxide (H_2O_2) 3% (v/v) was added to detect the catalase production by the bacterial isolates. If bubble formation was observed, it indicated the catalase-positive test.

- **Composition of TSA Medium:**

Pancreatic digest of casein	17.0 gl^{-1}
Enzymatic digest of soybean meal	3.0
Dextrose	2.50
NaCl	5.0
K_2HPO_4	2.50
Agar	20.0
pH	7.3

12.9.5.2 UREASE TEST

To detect the production of urease enzyme by the bacterial isolates, the urease test was performed. Bacterial culture of 24 h 0.1 ml was inoculated in 10 ml of sterile urea broth and kept at 37°C for 48 h. Positive isolates degraded urea present in the urea broth medium by means of urease enzyme. The turning of phenol red to deep pink indicated the presence of urease activity.

➢ **Composition of Urea Broth:**

Urea	20.0 gl^{-1}
Na_2HPO_4	9.50
KH_2PO_4	9.10
Yeast extract	0.10
Phenol red	0.01
pH	6.8

12.9.5.3 CITRATE TEST

The citrate test was performed to detect the production of citrate enzyme by the bacterial isolates. Simmons citrate agar 10 ml slants were prepared, inoculated with 24 h of bacterial culture and kept at 37°C for 48 h, followed by addition of bromothymol blue indicator over the surface. Positive isolates use citrate as a carbon source producing citrase enzyme. On use of citrate, the bacteria produce alkaline products by changing the color of the bromothymol blue from green (at neutral pH 6.9) to blue (at higher pH 7.6) in the medium.

➢ **Composition of Simmons Citrate Agar Slant Medium:**

Magnesium sulfate	0.20 gl^{-1}
Ammonium dihydrogen phosphate	1.0
Potassium phosphate	1.0
Sodium citrate	2.0
Sodium chloride	5.0
Bromothymol blue	0.08
Agar	15.00
pH	6.8

12.9.5.4 TRIPLE SUGAR IRON TEST

The test in tryptic nitrate broth with three sugars and iron was carried out to differentiate bacteria for their ability to ferment glucose, lactose, and sucrose or to reduce sulfur to hydrogen sulfide. A loopful of 24 h bacterial culture was streaked over TSI agar slants and kept at 37°C for 48 h. Changes in the color of the medium was observed over TSI agar slants.

- **Composition of Sugar:**

Casein enzymatic hydrolysate	10 gl^{-1}
Peptic digest of animal tissue	10
Yeast extract	3
Beef extract	3
Lactose	10
Sucrose	10
Dextrose	10
Ferric ammonium citrate	0.3
Sodium chloride	5.0

12.9.5.5 NITRATE REDUCTION TEST

The test conducted in the tryptic nitrate medium detects reduction of nitrate (NO_3) to nitrite (NO_2) or nitrogenous compounds like molecular nitrogen (N_2) using nitrate reductase enzyme. The nitrate medium contains potassium nitrate as the substrate. The bacterial culture of 24 h duration 1 ml was inoculated in 100 ml of sterile trypticase nitrate broth and kept at 37°C for 48 h. To confirm the nitrate reduction capacity of bacterial cultures, post 48 h of incubation, sulfanilic acid (Sol A), α-naphthylamine (Sol B)) and zinc powder were mixed with bacterial culture. If the organism reduces NO_3 to NO_2, the nitrite would react with sulfanilic acid and – naphthylamine to produce a cherry red color. If no color is developed, it indicates either of the following two reactions: (i) nitrate is not reduced or (ii) nitrate is reduced even further to compounds other than nitrite (NH_2 or N_2).

- **Composition of Tryptic Nitrate Medium:**

Casein enzymatic hydrolysate	10 gl^{-1}
Peptic digest of animal tissue	10
Yeast extract	3
Beef extract	3
Lactose	10

Sucrose	10
Dextrose	10
Ferric ammonium citrate	0.3
Sodium chloride	5.0
Sodium thiosulfate	0.3
Phenol red	0.024
Agar	12
pH	7.4

12.9.5.6 INDOLE PRODUCTION TEST

The test was used to identify bacteria able to produce indole using tryptophanase enzyme. Sterile sulfur indole motility (SIM) agar deep tubes were streaked with 24 h bacterial culture over agar surface and kept at 37°C for 48 h. The by-product indole was identified by adding Kovac's reagent having HCl, dimethyl-amino-benzaldehyde, and amyl alcohol, a red layer formation referred to the presence of indole.

12.9.5.7 H_2S PRODUCTION TEST

The test was used to identify sulfur reducing bacteria. In this test a loopful of 24 h bacterial culture was inoculated inside a 10 ml of sulfide indole motility medium (SIM) agar slant and kept at 37°C for 48 h. The medium contained cysteine, an amino acid containing sulfur, and sodium thiosulfate with peptonized iron or ferrous sulfate. Formation of black precipitate indicates that the test would be positive for hydrogen sulfide (H_2S) production by anaerobic respiration of bacterial culture. No precipitate formation would refer to a negative test.

➢ **Composition of the Sulfide Indole Motility Medium (SIM):**

Pancreatic digest of casein	20 gl^{-1}
Peptic digest of animal tissue	6.1
$Fe(NH_4)_2(SO_4)_2.6H_2O$	0.2
$Na_2S_2O_3.5H_2O$	0.2
Agar	3.5
pH	7.3

12.9.5.8 *LITMUS MILK REACTION TEST*

The test enables identification of bacteria able to transform enzymatically milk substrates to different metabolites. Sterile litmus milk broth 10 ml was inoculated with 0.1 ml of 24 h bacterial culture and kept at 37°C for 48 h. After the incubation, it was observed if any changes in medium color, lactose fermentation, gas formation, curd formation, litmus reduction, peptonization, and alkaline reaction occur.

➢ **Composition of Litmus Milk Medium:**

Skim milk powder	100 gl^{-1}
254/Litmus	0.5
Sodium sulfate	0.5
pH	6.8

12.9.5.9 *METHYL RED-VOGES-PROSKAUER (MR-VP) BROTH*

A methyl red (MR) test was carried out to test the ability of an organism to produce and maintain stable acid end products from the glucose fermentation. In this test, 100 ml of sterile MR-VP broth was inoculated with 1 ml of 24 h bacterial culture and kept at 37°C for 48 h. MR-VP medium was separated into parts A and B.

- In part A, the pH indicator MR was added in MR test for detecting the presence of high concentrations of end acid products. If the tube turns red in color, it indicates a positive test and if yellow it indicates a negative.
- In part B, mixtures of Barrit A and B solutions were added in VP broth. The Voges-Proskauer (VP) test was used to determine the production of non-acidic or neutral products, such as acetylmethylcarbinol (acetoin), a neutral product formed from pyruvic acid in the course of glucose fermentation. The reagent used in the VP test is Barrit's reagent which consists of a mixture of alcoholic–naphthol and 40% potassium hydroxide. After 15 minutes of addition of Barrit's reagent, deep rose color of the culture indicates the presence of acetoin which represented positivity.

➢ Composition of MR-VP Broth:

Buffered peptone	7 gl^{-1}
Dextrose	5
Dipotassium phosphate	5
pH	6.9

12.10 EXTRACELLULAR ENZYME ACTIVITY (HYDROLYSIS) TEST MEDIUM

- To determine the ability of microorganisms to excrete extracellular hydrolytic enzymes capable of degrading polysaccharides, lipids, and proteins (casein and gelatin), the hydrolysis test was carried out in the concerned medium. The test was used to differentiate microbes based on their ability to hydrolyze starch with exoenzyme amylase.
- For determination of hydrolytic activity, Tributyrin agar was used for the detection of lipase as produced by the concerned microorganisms. The nutrient agar supplemented with triglyceride tributyrin is used as the lipid source. In the experiment, milk agar was used to determine the hydrolytic activity of the enzyme. The positive isolates produce a clear zone surrounding their growth losing opacity. But the absence of protease activity, the medium surrounding the growth of the bacteria remained opaque, which was a negative reaction.
- Gelatin liquefaction test was used to determine the ability of bacteria to produce hydrolytic exoenzymes called gelatinases that digest and liquefy gelatin. The presence of these enzymes, as determined by the liquefaction, was used for identifying certain bacteria.

➢ Starch Agar Composition:

Peptone	**5 gl^{-1}**
Beef extract	3
Starch	2
Agar	15
pH	7.0
Tributyrin Agar Composition	
Peptone	5 gl^{-1}

Peptone	**5 gl^{-1}**
Yeast extract	3
Agar	15
pH	6.9
Milk Agar Composition	
Peptone	5 gl^{-1}
Skim milk powder	100
Agar	15
pH	6.9
Nutrient Gelatin Composition	
Peptone	5 gl^{-1}
Beef extract	3
Gelatin	120
pH	6.8

12.11 CARBOHYDRATE FERMENTATION MEDIUM

This biochemical test was performed to determine the ability of microorganisms to degrade and ferment carbohydrates with the production of acid and/or gas. For this purpose, different types of carbohydrate enriched media were used as presented in Table 12.5.

TABLE 12.5 Composition of Phenol Red Lactose, Dextrose, and Sucrose Broths

Chemicals	**Phenol Red Lactose**	**Phenol Red Dextrose**	**Phenol Red Sucrose**	**Phenol Red Xylose**	**Phenol Red Mannitol**
Protease peptone	10	10	10	10	10
Beef extract	1	1	1	1	1
Sodium chloride	5	5	5	5	5
Phenol red	0.018	0.018	0.018	0.018	0.018
Lactose	10	–	–	–	–
Dextrose	–	1	–	–	–
Sucrose	–	–	5	–	–
Xylose	–	–	–	5	–
Mannitol	–	–	–	–	5
pH	7.4	6.9	6.9	6.8	6.8

Fermented carbohydrates with the production of acidic wastes might cause yellowing of phenol from that of red color, thereby indicating a positive reaction. In some cases, acid production is accompanied by the evolution of CO_2 gas which becomes visible as bubbles in the inverted tube. Cultures that are not capable of fermenting carbohydrates are negative ones.

12.12 BIOMASS ESTIMATION

The bacterial biomass in the liquid culture media was estimated by the gravimetric method. The culture samples of the bacterial samples were taken in a polypropylene tube and centrifuged at 8,000 rpm for 15 min. The precipitated cells were washed with distilled water and dried at 60°C in a hot air oven until a constant weight was achieved.

12.13 POLYMER EXTRACTION AND PURIFICATION

For the extraction of PHA, biomass of the bacterial cells was harvested by centrifugation at 8,000×g from the culture broth. The pellet following washing twice with sterilized distilled water was lyophilized. PHA was extracted from the dry biomass by using Soxhlet extraction for 24 h using chloroform. The PHA was then concentrated by rotary vacuum evaporator and precipitated adding 10 volumes of ice-cold methanol. The process was trice repeated to get the purified polymer (Law and Slepecky, 1969), the polymer was collected by air drying.

12.13.1 QUANTIFICATION OF PHA BY GRAVIMETRIC METHOD

The PHA content in bacterial cells was determined gravimetrically (Steinbuchel, 2005). The PHA dissolved in chloroform obtained from Soxhlet extraction was washed with ice cold methanol to get purified product. The product was then collected by air drying, weighed, and PHA concentration was expressed as % of cellular dry weight (CDW).

$$\% \text{ PHA} = \text{Weight of PHA} \times 100/\text{Weight of biomass}$$

12.13.2 CHARACTERIZATION OF PHA BY UV-VISIBLE SPECTROPHOTOMETER ANALYSIS

The structure and stability of materials in solid and liquid forms can be highlighted by UV-Visible spectrophotometer. Different types of electronic or ionic excitation might occur in the molecules due to absorption of energies available in the Vis region. The spectrophotometer records the WL at the absorption course together with the degree of absorption. The spectrum thus derived is revealed graphically pertaining to absorbance vs WL. UV-Vis spectra were recorded on a Shimadzu UV-2500 UV-Vis spectrophotometer using chloroform.

12.13.3 FTIR ANALYSIS

FTIR characterization of the PHA containing cells and that of the purified PHA were done using Perkin Elmer, Spectrum 100 FTIR spectrometer to evaluate the functional groups of the compounds. Purified sample PHA was mixed with potassium bromide (KBr) using a clean and smooth mortar and pestle. After mixing for three minutes, the mass was transferred to a pellet preparation instrument, and a solid pellet was prepared using hydraulic pressure. The pellet was transferred to the holder of the FTIR spectrometer and the spectra thus obtained were recorded in the WL range of 400–4,000 cm^{-1}.

12.13.4 GC-MS ANALYSIS

PHA composition was analyzed by GC using either lyophilized cells or purified polymer samples. To determine the polymer content of the bacterial cells, freeze dried cell samples were subjected to methanolysis in the presence of H_2SO_4 according to Brandl et al. (1988). About 4 mg of bacterial cells was reacted with a solution containing 1 ml chloroform, 0.85 ml methanol and 0.15 ml H_2SO_4 for 140 min at 100°C in an oil bath. The intracellular PHA was degraded to 3-hydroxyalkanoic acid methyl esters. The reaction after adding 0.5 ml distilled water, the tube was shaked vigorously for 1 min. The bottom organic phase after separation was collected, dried over anhydrous sodium sulfate, filtered, and then analyzed (Barnard and Sanders, 1989). The methyl esters were analyzed by GC Varian 3800 model and Saturn 2,200 MS spectroscopy (Perkin Elmer) equipped with CP-Sil 8 CB and CP-Sil 5 CB capillary columns and flame ionization GC detector and quadruple ion trap MS detector. From the organic phase 2 μl was analyzed after split injection (ratio 1:40) and using helium (35 cm/min) carrier gas. Injection and detector temperature 230°C and 275°C, respectively was maintained. A temperature program for the efficient separation of methyl esters was used:

> 120°C for 5 min, with rise of 8°C/min, 15°C/min; P(HB) and P(HB-co-HV) were used as standards; benzoic acid was used as an internal standard.

12.14 NMR ANALYSIS

To confirm the chemical structure of PHA, 1H and 13C NMR analyzes were carried out. The spectra were measured with JEOL JNN-ECS 400 spectrophotometer. A sample of 20 mg purified polymer was dissolved in 1 mL deuterated chloroform ($CDCl_3$) and analyzed with 1H NMR (400 MHz) and 13C NMR (400 MHz) at 25°C. The signal of tetramethylsilane related to the chemical shifts in ppm was reported.

12.15 CRYSTALLINITY STUDY BY X-RAY DIFFRACTOGRAM (XRD)

XRD fraction measurement was done with a Miniflex Table Top XRD (model Rigaku Corporation) and a Ni filter. The X-ray diffractogram obtained from the polyester pellet samples was aged at room temperature to reach crystallinity equilibrium prior to analysis and recorded at 27°C in the range 2 = 5°–70° at a scan rate of 0.01/10s. The crystallinity degree (Xc) was calculated from normalized diffractogram data recorded from the relation:

$$Xc = [(\text{Total area}) - (\text{Amorphous area})]/(\text{Total area})$$

12.16 SURFACE STUDY BY SEM

A microstructural study for the surface topography of the isolated polymer was done by using a JEOL JSM Model 6390 LV Asia PTE Ltd, Singapore.

Samples were placed in aluminum stubs of 8 mm diameter and then coated with platinum using the sputtering device. The images at the acceleration voltage of 15 kV (max) were taken to avoid polymer incineration due to the beam heat.

12.17 DETERMINATION OF MOLECULAR WEIGHT (MW) BY GEL PERMEATION CHROMATOGRAPHY (GPC)

GPC (Waters GPC system) was used to determine the mol wt. of the isolated polymer. The PHA was dissolved in THF (tetrahydrofuran) and analyzed. A polystyrene standard with a low polydispersity was used in generating calibration curve equipped with serially connected RI detectors. No further corrections in THF were used as the eluent at the flow rate of 0.5 $mL.min^{-1}$ and temperature 40°C o determine the mol wt.

12.18 THERMAL PROPERTY BY DSC ANALYSIS

DSC data of the polymer sample (3 mg) encapsulated with aluminum pans, heating at–30 to 300°C was recorded at the rate of 20°$C.min^{-1}$ (Shimadzu DSC-50 system) equipped with a cooling accessory under the nitrogen flow of 30 $ml.min^{-1}$. The testing temperature was maintained at 150°C for 1 min and quenched to 30°C at a rate of 20°$C.min^{-1}$. Subsequently, the polymer sample was reheated from 30°C to 300°C at a heating rate of 20°$C.min^{-1}$. The midpoint of increasing step was taken as the glass transition temperature (Tg). Upon heating an endothermic peak was observed, and the temperature at the peak was the melting temperature (Tm).

12.19 BY TGA ANALYSIS

Polymer thermal stability and decomposition profile were determined by using Thermogravimetric analyzer (Shimadzu TGA-50) functioning at a nitrogen flow rate of 10 $ml.min^{-1}$. The sample was weighed to *ca.* 10 mg and placed in a platinum pan and heated from 30–600°C at 10°$C.min^{-1}$ rate under the nitrogen atmosphere. The decomposition temperature to the maximum was recorded from the thermograms and determined by analysis of the derivative weight loss curve over temperature.

12.20 PL STUDY

Photoluminescence (PL) spectroscopy is used to analyze fluorescent compounds at very low concentration in both solid and liquid states. Ions or molecules show fluorescence spectra when they absorb electromagnetic radiation at higher energy short wavelengths and are capable of radiating at lower energy longer wavelengths. PL spectra were recorded using a PerkinElmer model LS 55 spectrophotometer, using excitation of the polymer at maximum absorption wavelength. The PL spectrum of the polymer solution was recorded in the chloroform solvent. PL of solid-state polymers were measured by preparing the polymer film on the glass substrate. The emission was recorded at two different wavelengths for all the polymers.

12.21 SCREENING OF BIOPOLYMER PRODUCING MICROBES

12.21.1 ISOLATION AND PURE CULTURE OF MICROBES CAPABLE OF PRODUCING BIOPOLYMER

A good number of culturable bacteria were obtained from the soil samples of crude oil contaminated sites. A total of 13 bacterial isolates were screened based on their ability to produce biopolymer. The bacterial isolates were recovered, pure cultured and maintained in stab agar cultures at 4°C and preserved in 15% glycerol at–80°C.

The bacterial isolates were cultured in NB medium at 37°C in an incubator shaker at 180 rpm for 24 h, and subsequently the pure cultures were used as the seed culture for further screening. The pure cultures were sub-cultured by re-suspending the microbes in fresh polymer detection media with the necessary environmental conditions.

12.21.2 MORPHOLOGICAL CHARACTERIZATION OF BACTERIAL ISOLATES

The morphology of the isolated and pure cultured bacterial strains was studied in the nutrient agar medium. Data thus obtained are presented in Table 12.6. Morphological study of bacteria in terms of colony size, pigmentation, form, margin, and elevation displayed variable results.

TABLE 12.6 Morphological Characters of Bacterial Isolates Obtained from Different Soil and Waste Samples

SL. No.	Bacterial Isolate	Size	Form	Margin	Elevation	Pigment	Gram Staining
1.	BS1	Small	Circular	Entire	Flat	Greenish	Gram –ve
2.	BS2	Medium	Circular	Entire	Convex	Yellowish	Gram –ve
3.	BP C1	Large	Circular	Entire	Flat	Bluish-green	Gram –ve
4.	BP C2	Large	Circular	Entire	Flat	Greenish	Gram –ve
5.	BS4	Medium	Circular	Entire	Flat	Yellowish green	Gram –ve
6.	BS5	Large	Circular	Entire	Raised	Greenish	Gram –ve
7.	BS7	Small	Circular	Entire	Flat	Yellowish green	Gram –ve
8.	BP1	Medium	Circular	Entire	Convex	White	Gram –ve
9.	BP2	Small	Circular	Serrate	Flat	White	Gram +ve
10.	BP3	Pinhead	Circular	Entire	Flat	White	Gram –ve
11.	BP4	Large	Circular	Entire	Raised	White	Gram +ve
12.	BPr2	Large	Circular	Entire	Flat	White	Gram –ve
13.	BPr3	Medium	Circular	Entire	Flat	White	Gram +ve

12.21.3 SCREENING FOR PHA PRODUCING BACTERIA

In screening experiments, all the bacterial isolates were cultured in the NB (Sigma, Aldrich) medium at 37°C as a pre-culture for 24 h. The PHA detection medium (Phukon et al., 2014) was used with minor modifications to produce PHA. A pH of 7 in the medium was adjusted prior to autoclaving. The composition of the medium (g.l^{-1}) is: $(NH_4)2SO_4$ (1.0), KH_2PO_4 (13.3), $MgSO_4$ (1.3), citric acid (1.7), trace element solution 10 mL^{-1} (g.l^{-1}, $FeSO_4 \cdot 7H_2O$ 10, $ZnSO_4 \cdot 7H_2O_2$ 0.25, $CuSO_4 \cdot 5H_2O$ 1, $MnSO_4 \cdot 5H_2O$ 0.5, $CaCl_2 \cdot 2H_2O$ 2.0, $Na_2B_4O_7 \cdot 10H_2O$ 0.23, $(NH_4)6Mo_7O_{24}$ 0.1, and HCl 10 ml. To avoid reduction of glucose by high temperature and pressure, it was added to the medium after autoclaving. An aliquot of 10 ml of the above pre-cultured inoculum was added to 100 ml sterile PHA medium containing glucose as the carbon source, cultured in a shaker running at 180 rpm for 48 h and temperature 37°C.

Bacterial multiplication was indicated by the increase in the turbidity of the culture medium. The bacterial biomass was harvested by centrifugation and then assessed for PHA production. The screening data are presented in Table 12.7.

TABLE 12.7 Screening of Biopolymer Producing Bacteria

Isolate No.	Response in PHA Medium
BS1	–
BS2	–
BP C1	+
BP C2	+
BS4	–
BS5	–
BS7	–
BP1	–
BP2	+
BP3	–
BP4	–
BPr2	–
BPr3	–

As shown in Table 12.7, out of 13 pure bacterial cultures, only three such as BPC1, BPC2, and BP2 showed growth in the PHA detection medium. These isolates were promising to produce biopolymer PHA. Subsequent studies were carried out with these 3 bacterial isolates.

BPC1 and BPC2 bacterial isolates were identified by assaying the 16S-rRNA and the same were deposited in GenBank NCBI and the accession No given are *P. aeruginosa* BPC1 (JQ796859) and BPC2 (JQ866912), respectively. Figure 12.3 Phylogenetic tree the details of the genotypic characterization of the two strains are presented in the subsequent section. The strain BP2 was identified as *Bacillus circulans* MTCC8167 in a previous work in our laboratory. Earlier the strain BP2 was examined for biosurfactant production but not for biopolymer production (Bordoloi and Konwar, 2009).

12.21.4 SCREENING OF PHA PRODUCING BACTERIAL ISOLATE BY STAINING PROCEDURE

The presence of biopolymer in the growing bacterial cells was confirmed by the Fluorescence Staining as described by Ostle and Holt (1982). Following the staining, the bacterial cultures were observed under a fluorescence microscope with fluorescence excitation using immersion oil attachment. The PHA producing bacteria were also stained by Sudan black B as per the method of Schlegel et al. (1970) for visualizing the PHA granules in the bacterial cells under a light microscope. The observation is presented in Table 12.8.

TABLE 12.8 PHA Producing Bacterial Isolates Showing Staining Response

Strains	Nile Blue A Treatment	Sudan Black B Treatment
BS1	–ve	–ve
BS2	–ve	–ve
P. aeruginosa JQ796859	+ve	+ve
P. aeruginosa JQ866912	+ve	+ve
BS4	–ve	–ve
BS5	–ve	–ve
BS7	–ve	–ve
BP1	–ve	–ve
B. circulans MTCC8167	+ve	+ve
BP3	–ve	–ve
BP4	–ve	–ve
BPr2	–ve	–ve
BPr3	–ve	–ve

As shown in Table 12.8, out of 13 bacterial isolates, only three such as *P. aeruginosa* JQ796859 and JQ866912, *B. circulans* MTCC8167 showed orange fluorescence after the Nile blue A treatment (Figure 12.4). Similarly, when Sudan black B (Figure 12.5) was applied, the PHA rich bacterial cells retained the dye and appeared dark blue under the microscope. Colonies of PHA deficient cells appeared light gray as they lost the dye during the differentiation process. The positive isolates were promising to produce biopolymer; PHA.

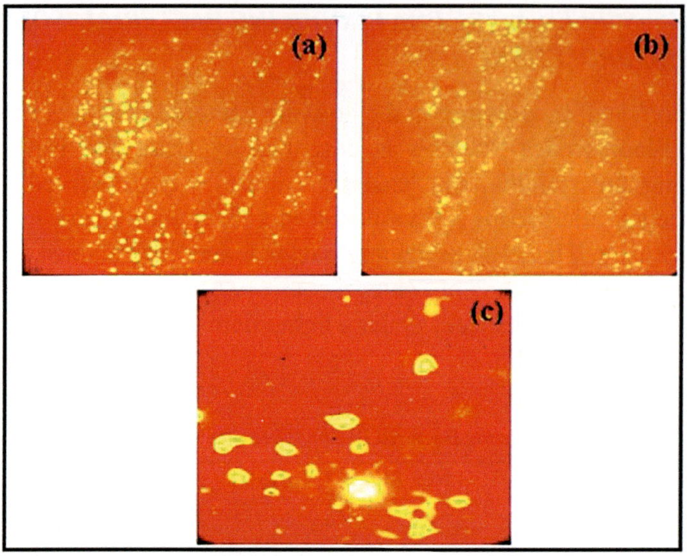

FIGURE 12.4 PHA inclusion bodies in bacterial isolate. (a) *P. aeruginosa* JQ796859; (b) *B. circulans* MTCC8167; and (c) *P. aeruginosa* JQ866912 stained with Nile blue A and observed under fluorescent light (100X magnification).

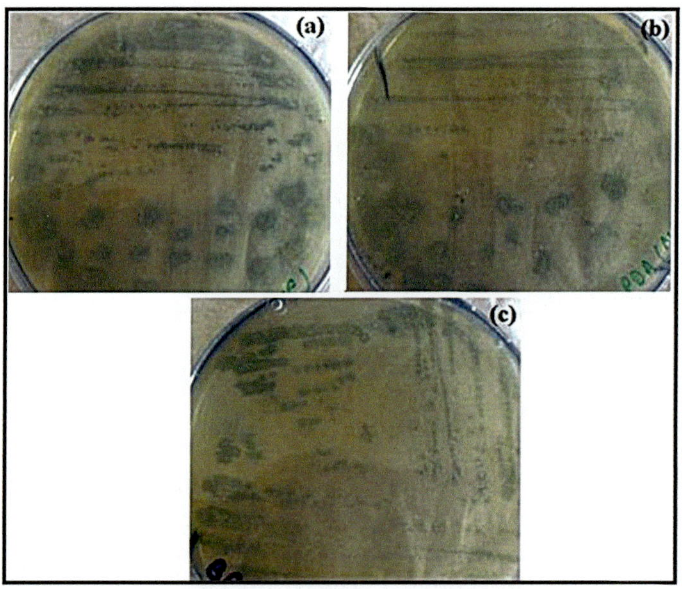

FIGURE 12.5 Sudan black B treatment on bacterial isolates. (a) *P. aeruginosa* JQ796859; (b) *B. circulans* MTCC8167; and (c) *P. aeruginosa* JQ866912. Positive isolates showing bluish black color of the colonies.

KEYWORDS

- **biopolymer**
- **minimum salt medium**
- **Nile blue A treatment**
- **polyhydroxyalkanoates**
- **scanning electron micrograph**
- **sulfur indole motility**
- **trypticase soy agar**

REFERENCES

Barnard, G. N., & Sanders, J. K., (1989). The poly-beta-hydroxybutyrate granule *in vivo*. A new insight based on NMR spectroscopy of whole cells. *J. Biol. Chem., 264*, 3286–3291.

Bordoloi, N. K., & Konwar, B. K., (2008). Microbial surfactant-enhanced mineral oil recovery under laboratory conditions. *Colloids and Surfaces B: Biointerfaces, 63*, 73–82.

Bordoloi, N. K., & Konwar, B. K., (2009). Bacterial biosurfactant in enhancing solubility and metabolism of petroleum hydrocarbons. *J. Hazardous Materials, 170*, 495–505.

Brandl, H., et al., (1988). *Pseudomonas oleovorans* as a source of poly (β-hydroxyalkanoates) for potential applications as biodegradable polyesters. *Appl. Environ. Microbiol., 54*, 1977–1982.

Cappuccino, G. J., & Sherman, N., (1999). *Microbiology a Laboratory Manual* (4th edn., Vol. 37). Wesley. ISBN 0-201-52628-X: 129-183.

Haas, R., et al., (2008). Production of poly (3-hydroxybutyrate) from waste potato starch. *Biosci. Biotechnol. Biochem., 72*, 253–256.

Jamil, N., et al., (2007). Characterization of biopolymer produced by *Pseudomonas* sp. CMG607w of marine origin. *J. Gen. Appl. Microbiol., 53*, 105–109.

Ostle, A. G., & Holt, J. G., (1982). Nile Blue A as a fluorescent stain for poly-b-hydroxybutyrate. *Appl. Environ. Microbiol., 44*, 238–241.

Phukan, P., Saikia, J. P., & Konwar, B. K., (2011). Bioplastic (P-3HB-co-3HV) from *Bacillus circulans* (MTCC 8167) and its biodegradation. *Colloids and Surfaces B: Biointerfaces, 92*, 30–34.

Phukon, P., Phukan, M. M., Phukan, S., & Konwar, B. K., (2014). Polyhydroxyalkanoate production by indigenously isolated *Pseudomonas aeruginosa* using glycerol by-product of KCDL biodiesel as an inexpensive carbon source. *Annals of Microbiology, 64*(4), 1567–1574.

Qi, Q., et al., (1997). Synthesis of poly (3-hydroxyalkanoates) in *Escherichia coli* expressing the PHA synthase gene phaC2 from *Pseudomonas aeruginosa*: Comparison of PhaC1 and PhaC2. *FEMS Microbiol. Lett., 157*, 155–162.

Rehm, B. H. A., (2003). Polyester synthases: Natural catalysts for plastics. *Biochem. J., 376*, 15–33.

Rehman, S., et al., (2007). Screening of different contaminated environments for polyhydroxyalkanoates-producing bacterial strains. *Biol Bratislava, 62*, 650–656.

Saitou, N., & Nei, M., (1987). The neighbor-joining method: A new method for reconstructing phylogenetic trees. *Mol Bio. Evol., 4*, 406–425.

Schlegel, H. G., et al., (1970). The isolation of mutants not accumulating poly-b-hydroxybutyric acid. *Arch Microbiol., 71*, 283–294.

Solaiman, D. K, Y., et al., (2000). Rapid and specific identification of medium-chain-length polyhydroxy alkanoates synthase gene by polymerase chain reaction. *Appl. Microbiol. Biotechnol., 53*, 690–694.

Solaiman, D. K. Y., (2002a). PCR cloning of *Pseudomonas resinovorans* polyhydroxyalkanoates biosynthe sis genes and expression in *Escherichia coli*. *Biotechnol. Lett., 22*, 789–794.

Solaiman, D. K. Y., (2002b). Polymerase-chain-reaction-based detection of individual polyhydroxyalkanoates synthase phaC1 and phaC2 genes. *Biotechnol. Lett., 24*, 245–250.

Steinbuchel, A., (2005). Non-biodegradable biopolymers from renewable resources: Perspectives and impacts. *Curr. Opin. Biotechnol., 16*, 607–613.

Zhang, G., et al., (2001). PCR cloning of type II polyhydroxyalkanoate biosynthesis genes from two *Pseudomonads*. *FEMS Microbiol. Lett., 198*, 165–170.

CHAPTER 13

In Vitro Biodegradation Study of PHA Film by Soil Microorganisms

The extracted polymer from the isolated bacteria was dissolved in chloroform in a specific amount and cast over a cover slip (2 cm^2) homogeneously. The film was kept for 24–36 h at room temperature undisturbed to get a thin film of the PHA polymer.

In the present investigation, bacteria and fungi species isolated from crude oil contaminated soils of Assam were used to degrade the extracted polymer. Bacteria used were *Pseudomonas aeruginosa* strains BP4, BP5, BP7, *Bacillus subtilis* R38-I, *B. circulans* MTCC8167, *Alcaligens faecalis* MTCC8164, *P. aeruginosa* MTCC7815, *Mycobacterium* spp. G-35I and two fungal strains *Candida albicans* MTCC3017 and *Fusarium oxysporum* NCIM1281. Bacteria and fungi strains used were maintained on the Nutrient agar and Sabouraud dextrose (SD) agar slants, respectively. Mineral salt medium used in the investigation was composed of KNO_3, 2; K_2HPO_4, 1; KH_2PO_4, 0.7; $MgSO_4 \cdot 7H_2O$, 0.7 g/l in distilled water and the pH adjusted to 7. Accordingly, fungal species were grown on SD medium for the polymer degradation test. Aliquots (10 ml) of the mineral salt medium and SD medium were dispensed in 100 ml conical flasks, each inoculated with 5 pieces of PHA films (2 cm^2). Flasks were incubated at 37°C and 30°C for bacteria and fungi, respectively for 36 days. The experiment was performed thrice. Two flasks containing mineral salt medium and SD medium separately with no PHA film were used as the control for the bacterial and the fungal species, respectively. After completion of incubation, the PHA films were taken out and washed with

Bacterial Biopolymers. B. K. Konwar, PhD (Author)

sterile distilled water. The effect of microbial treatment was monitored by SEM and FTIR.

13.1 SEM OBSERVATION

The surface topology of the PHA films, incubated in sterile fluid for 36 days and those inoculated with bacteria and fungi were analyzed through optical microscope (JEOL JSM Model 6390 LV Asia PTE Ltd. Singapore) to verify for any structural change. Images obtained by the SEM analysis on the treated and untreated samples were analyzed on a comparative basis.

13.2 FTIR SPECTROSCOPY ANALYSIS

FTIR (PerkinElmer, Spectrum 100) analysis was done to detect the degradation of PHA. Part of the polymer film was mixed with KBr and made into a pellet, which was fixed to the FTIR sample plate. Spectra were recorded in triplicate (400–4,000 cm^{-1}) for each sample.

13.3 MOLECULAR BIOLOGY OF PHA PRODUCING BACTERIA

13.3.1 ISOLATION AND PURIFICATION OF DNA

Genomic DNA from bacteria was prepared as described by Sambrrok et al. (2001). Cells were pelleted by spinning 4 ml of fresh bacterial culture at 9,450 × g for 10 min in a refrigerated Beckman Centrifuge and the supernatant was decanted and discarded. The pellet was dissolved in 0.8 ml of solution I (50 mM glucose, 25 mM Tris cl pH 8.0, 10 mM EDTA pH 8.0) and vortexed. To this resultant mixture, 160 µl of lysozyme (10 $mg.ml^{-1}$) was added and incubated at room temperature (24°C) for 20 min. Subsequently, 44.5 µl of 10% (w/v) SDS solution was added and re-incubated for 10 min at 50°C. After that, 53.3 µl of RNAase (10 $mg.ml^{-1}$) was added to the above sample and incubated at 37°C for 90 min. This was followed by addition of 45.3 µl of Na-EDTA (0.1 M, pH 8) and re-incubated at 50°C for 10 min. To remove the protein, 26.6 µl of proteinase K (5 $mg.ml^{-1}$ stock) was added and incubated at 50°C for 10 min. An equal volume of phenol (saturated with 0.1 M Tris HCl, pH

8) was added to the above solution and mixed thoroughly. The mixture was centrifuged at 9,450 × g for 10 min, the upper aqueous phase was aspirated into the sterile Eppendorf tubes whereas the lower phase was discarded. Equal volume of [Phenol: chloroform (1:1)]: isoamyl alcohol (24:1) was added and mixed thoroughly. Following centrifugation at 9,450 × g for 10 min, the upper phase was transferred to a sterile microfuge tube, and then an equal volume of chloroform-isoamyl alcohol (24:1) was added and spun at 9,450 × g for 10 min. The upper phase was transferred to a sterile microfuge tube and $1/10^{th}$ volume of Na-acetate (3M, pH 7) was added. The DNA was precipitated by adding 2 volumes of ice-cold absolute ethanol in the above solution and the DNA pellet recovered by centrifugation. After removal of the alcohol, DNA was resuspended in 10 mM Tris HCl^{-1} mM EDTA buffer (pH 8) at a final concentration of 1 $\mu g.ml^{-1}$ and stored at 4°C for further use.

13.3.2 GEL ELECTROPHORESIS

- **Reagents:** Agarose gel 0.8%, 50X and 1X TAE buffer, loading dye and ethidium bromide solution.
- **Procedure:**
  - Agarose gel 0.8% was prepared with 2% 50X TAE buffer and poured into the gel caster;
  - Allowed the gel to solidify;
  - The gel was placed in the electrophoresis chamber filled with 1X TAE buffer;
  - DNA sample preparation: DNA samples 5–10 µl was mixed with 2 µl of loading dye;
  - The sample was run at 100 volts till the loading dye reached 75% of the gel;
  - The gel was removed from the electrophoresis chamber and examined on a UV transilluminator.

13.3.3 PCR AMPLIFICATION OF 16S-RRNA GENE

DNA amplification was performed with Genamp PCR system (Applied Biosystem, USA). Reaction mixture for the PCR contained 10x PCR buffer [200 mM Tris-HCl (pH 8.4), 500 mM KCl], each deoxynucleotide triphosphate (dNTP) at a final concentration of 2.5 mM, 1.5 mM $MgCl_2$, 400 ng of each oligonucleotide primer and 3 U Taq DNA polymerase (Invitrogen, USA) in a final volume of 100 μl. The PCR product was sequenced bi-directionally using the forward, reverse, and internal primer of the 16S-rDNA. The sequences of the universal primers are presented in Table 13.1.

TABLE 13.1 Bidirectional Sequencing with Forward, Reverse, and Internal Primer of the 16S-rDNA

SL. No.	Primers
1.	16S Forward primer (27F): (5'-GAGTRTGATCMTYGCTWAC-3'):19 Mer
2.	16S Reverse primer (1544 R): (5'-CGYTAMCTTWTTACGRCT-3'):18 Mer
3.	Internal primer (13BG): (5'-CAGCAGCCGCGGTAATAC-3'):18 Mer

The PCR condition was standardized with respect to each strain as shown in Table 13.2. The amplified DNA was verified by electrophoresis of aliquots of PCR product (5 μl) on a 1% agarose gel in 1X TAE (Tris-acetate-EDTA) buffer. The PCR products (16S-rDNA) were purified using a gel extraction kit (Bangalore Genei) and sequenced with Big Dye Terminator version 3.1//Cycle sequencing kit and ABI 3500 XL Genetic Analyzer. The sequence data were aligned and compared with the published sequences obtained from the GenBank database using Seq Scape v 5.2.

TABLE 13.2 Optimal PCR Reaction Conditions for Amplification of Conserved Region of 16S-rRNA Gene of Selected Polymer Producing Bacterial Strains

PCR Conditions	Bacterial Strains	
	***P. aeruginosa* JQ796859**	***P. aeruginosa* JQ866912**
Initial denaturation	94°C for 5 min	94°C for 5 min
Denaturation	94°C for 30 s	94°C for 30 s
Annealing	55°C for 30 s	55°C for 30 s
Elongation	72°C for 2 min	72°C for 2 min
	72°C for 5 min	72°C for 5 min
Final elongation	72°C for 5 min	72°C for 5 min
Cycles	35	35

The sequencing mix reaction was carried out using:

➤ Composition of the sequencing mix (10 μl):

  - o Big dye terminator ready reaction mix: 4 μl;
  - o Template (100 ng/μl): 1 μl;
  - o Primer (10 pmol/λ): 2 μl;
  - o Milli Q water: 3 μl.
- ➤ PCR conditions for 25 cycles:
  - o Initial denaturation: 96°C for 1 min;
  - o Denaturation: 96°C for 10 sec;
  - o Hybridization: 50°C for 5 sec;
  - o Elongation: 60°C for 4 min;
  - o Phylogenetic analysis.

The 16S-rDNA sequence of the bacteria under study was aligned with the reference sequences showing homology from the NCBI database using the multiple sequence alignment program of MEGA4. A phylogenetic tree was constructed using ClustalW by distance matrix analysis and the neighbor-joining method. Phylogenetic trees were displayed using TREEVIEW (Dubnau et al., 1965; Vaneechoutte et al., 2004; Pohlmann et al., 2006). The 16S-rDNA gene sequence determined in this study was deposited in GenBank of NCBI data library under different accession number with respect to each strain.

13.3.4 PCR-BASED IDENTIFICATION OF THE PHA BIOSYNTHETIC GENES IN BACTERIAL STRAINS

Genomic DNA was isolated from bacterial strains *P. aeruginosa* strains JQ796859 and JQ866912 by using Genei Pure TM Bacterial DNA purification kit (Cat No: 117290). The primers used in the experiment were I-179L and I-179R (Vaneechoutte et al., 2004); custom-ordered from Bangalore Genei were based on two highly conserved sequences deduced from a multiple alignment analysis of the pseudomonas PhaC genes. The sequences of the primers are presented in Table 13.3.

TABLE 13.3 Sequence of I-179L and I-179R Primers

SL. No.	Primers
1.	Forward: I-179L–5′-ACAGATCAACAAGTTCTACATCTTCGAC-3′):28 Mer
2.	Reverse: I-179R–5′-GGTGTTGTCGTTGTTCCAGTAGAGGATGTC-3′):30 Mer

The PCR mixture consisted of 2 μl of the genomic DNA template; 0.45 μl (0.3 μM) of each primer, 1.5 μl of 1x PCR buffer, 0.6 μl of 200 μM dNTPs, 0.3 μl of (1U) Taq DNA polymerase in a final volume mixture of 15 μl. PCR amplification was performed with a Gene Amp PCR System 9700 (Applied Biosystems). The thermal cycling was initiated by denaturing the DNA at 95C for 2 min followed by 30 cycles of 94°C for 30 sec, 60°C for 45 sec and 72°C for 45 sec, with a final 7 min extension step at 72°C.

A portion of the amplified PCR reaction mixture with ethidium bromide was loaded on an agarose gel and separated by electrophoresis in TAE buffer (Tris 1.6 M, acetic acid 0.8 M, EDTA 40 mM). DNA ladder 100 bp (Bangalore Genei, India) was used as a molecular marker and loaded in the gel. After electrophoretic separation, PCR products were analyzed by UV transilluminator.

13.4 MOLECULAR GENETIC ASSESSMENT OF PHA PRODUCING BACTERIAL ISOLATES

13.4.1 PHYLOGENETIC ANALYSIS ON THE SEQUENCE OF THE 16S-RDNA CONSERVED REGION

The genomic DNA isolated from the bacterial isolates BP C1 and BP C2 was used for the amplification of 16S-rDNA by PCR. The 16S-rRNA genes were amplified and sequenced, and the sequences are shown in Figures 13.1 and 13.2.

13.4.2 PHYLOGENY OF STRAIN BP C1

A homologous search result of the bacterial strain BP C1 demonstrated 98–99% similarity of 16S-rDNA sequence with other species of the genus *Pseudomonas* as shown in Table 13.4. The phylogenetic tree constructed from the sequence data by the neighbor-joining method showed that *Pseudomonas aeruginosa* P21 (HQ697283) showing 99% homology represented the closest phylogenetic neighbor of the strain BP C1 (Figures 13.3 and 13.4).

```
CGGGGGGCAGGCCTACACATGCAGTCGAGCGGATGAAGGGAGCTTGCTCCTGGATTCAGCGGCGGACGGGTGA
GTAATGCCTAGGAATCTGCCTGGTAGTGGGGGATAACGTCCGGAAACGGGCGCTAATACCGCATACGTCCTGA
GGGAGAAAGTGGGGGATCTTCGGACCTCACGCTATCAGATGAGCCTAGGTCGGATTAGCTAGTTGGTGGGGTA
AAGGCCTACCAAGGCGACGATCCGTAACTGGTCTGAGAGGATGATCAGTCACACTGGAACTGAGACACGGTCC
AGACTCCTACGGGAGGCAGCAGTGGGGAATATTGGACAATGGGCGAAAGCCTGATCCAGCCATGCCGCGTGTG
TGAAGAAGGTCTTCGGATTGTAAAGCACTTTAAGTTGGGAGGAAGGGCAGTAAGTTAATACCTTGCTGTTTTG
ACGTTACCAACAGAATAAGCACCGGCTAACTTCGTGCCAGCAGCCGCGGTAATACGAAGGGTGCAAGCGTTAA
TCGGAATTACTGGGCGTAAAGCGCGCGTAGGTGGTTCAGCAAGTTGGATGTGAAATCCCCGGGCTCAACCTGG
GAACTGCATCCAAAACTACTGAGCTAGAGTACGGTAGAGGGTGGTGGAATTTCCTGTGTAGCGGTGAAATGCG
TAGATATAGGAAGGAACACCAGTGGCGAAGGCGACCACCTGGACTGATACTGACACTGAGGTGCGAAAGCGTG
GGGAGCAAACAGGATTAGATACCCTGGTAGTCCACGCCGTAAACGATGTCGACTAGCCGTTGGGATCCTTGAG
ATCTTAGTGGCGCAGCTAACGCGATAAGTCGACCGCCTGGGGAGTACGGCCGCAAGGTTAAAACTCAAATGAA
TTGACGGGGGCCCGCACAAGCGGTGGAGCATGTGGTTTAATTCGAAGCAACGCGAAGAACCTTACCTGGCCTT
GACATGCTGAGAACTTTCCAGAGATGGATTGGTGCCTTCGGGAACTCAGACACAGGTGCTGCATGGCTGTCGT
CAGCTCGTGTCGTGAGATGTTGGGTTAAGTCCCGTAACGAGCGCAACCCTTGTCCTTAGTTACCAGCACCTCG
GGTGGGCACTCTAAGGAGACTGCCGGTGACAAACCGGAGGAAGGTGGGGATGACGTCAAGTCATCATGGCCCT
TACGGCCAGGGCTACACACGTGCTACAATGGTCGGTACAAAGGGTTGCCAAGCCGCGAGGTGGAGCTAATCCC
ATAAAACCGATCGTAGTCCGGATCGCAGTTTGCATCTAGCTTGCGGGAAGTCGGAATCGCTAGTAATCGCCAA
TCAGAATGTCACGGTGAATACGTTCCCGGACCTTGTACACACCGCCCGTCACACCATGGGAGTGGGTTTCTCC
AGAAGTAGCTAGTACAACCGCAAGGGGGACGGTACCACGGAGTGATTCATGACTGGGGTGAAGTCGAACAACG
AGCCAAGCTGT
```

FIGURE 13.1 Partial DNA sequence of conserved region of 16S-rRNA gene of the PHA producing bacterial isolate BP C1.

```
CGGTGGCGGCGGCCTAACACATGCAAGTCGAGCGGATGAAGGGAGCTTGCTCCTGGATTCAGCG
GCGGACGGGTGAGTAATGCCTAGGAATCTGCCTGGTAGTGGGGGATAACGTCCGGAAACGGGCG
CTAATACCGCATACGTCCTGAGGGAGAAAGTGGGGGATCTTCGGACCTCACGCTATCAGATGAGC
CTAGGTCGGATTAGCTAGTTGGTGGGGTAAAGGCCTACCAAGGCGACGATCCGTAACTGGTCTGA
GAGGATGATCAGTCACACTGGAACTGAGACACGGTCCAGACTCCTACGGGAGGCAGCAGTGGGG
AATATTGGACAATGGGCGAAAGCCTGATCCAGCCATGCCGCGTGTGTGAAGAAGGTCTTCGGATT
GTAAAGCACTTTAAGTTGGGAGGAAGGGCAGTAAGTTAATACCTTGCTGTTTTGACGTTACCAACA
GAATAAGCACCGGCTAACTTCGTGCCAGCAGCCGCGGTAATACGAAGGGTGCAAGCGTTAATCGG
AATTACTGGGCGTAAAGCGCGCGTAGGTGGTTCAGCAAGTTGGATGTGAAATCCCCGGGCTCAAC
CTGGGAACTGCATCCAAAACTACTGAGCTAGAGTACGGTAGAGGGTGGTGGAATTTCCTGTGTAG
CGGTGAAATGCGTAGATATAGGAAGGAACACCAGTGGCGAAGGCGACCACCTGGACTGATACTGA
CACTGAGGTGCGAAAGCGTGGGGAGCAAACAGGATTAGATACCCTGGTAGTCCACGCCGTAACGA
TGTCGACTAGCCGTTGGGATCCTTGAGATCTTAGTGGCGCAGCTAACGCGATAAGTCGACCGCCT
GGGGAGTACGGCCGCAAGGTTAAAACTCAAATGAATTGACGGGGGCCCGCACAAGCGGTGGAGC
ATGTGGTTTAATTCGAAGCAACGCGAAGAACCTTACCTGGCCTTGACATGCTGAGAACTTTCCAGA
GATGGATTGGTGCCTTCGGGAACTCAGACACAGGTGCTGCATGGCTGTCGTCAGCTCGTGTCGTG
AGATGTTGGGTTAAGTCCCGTAACGAGCGCAACCCTTGTCCTTAGTTACCAGCACCTCGGGTGGG
CACTCTAAGGAGACTGCCGGTGACAAACCGGAGGAAGGTGGGGATGACGTCAAGTCATCATGGCC
CTTACGGCCAGGGCTACACACGTGCTACAATGGTCGGTACAAAGGGTTGCCAAGCCGCGAGGTGG
AGCTAATCCCATAAAACCGATCGTAGTCCGGATCGCAGTCTGCAACTCGACTGCGTGAAGTCGGA
ATCGCTAGTAATCGTGAATCAGAATGTCACGGTGAATACGTTCCCGGGCCTTGTACACACCGCCC
GTCACACCATGGGAGTGGGTTGCTCCAGAAGTAGCTAGTCTAACCGCAAGGGGGACGGTACCTCT
GGAGGTTTCCCT
```

FIGURE 13.2 Partial DNA sequence of conserved region of 16S-rRNA gene of the PHA producing bacterial isolates BP C2.

TABLE 13.4 Homologous Search Results of Partial Sequence of 16S-rRNA Gene of Strain BP C1 Using BLAST from NCBI*

Accession	Bacterial Species	S_ab Score
EU147007	*Pseudomonas putida* CGL-3	0.994
EU826025	*P. aeruginosa* CDB35	0.997
GQ472194	*Pseudomonas* sp. XQ-ph1	0.995
EU449119	*Pseudomonas* sp. 9 GUW	0.995
HM355700	*P. aeruginosa* BAC3111	0.994
HM355726	*P. aeruginosa* BAC1137	0.994
HQ697283	*P. aeruginosa* P21	1.000
HQ697284	*P. aeruginosa* P22	1.000
HQ697285	*P. aeruginosa* P28	0.998
HQ697286	*P. aeruginosa* P43	0.998

*The 16S-rDNA sequences from microbes showing up to 99% identity.

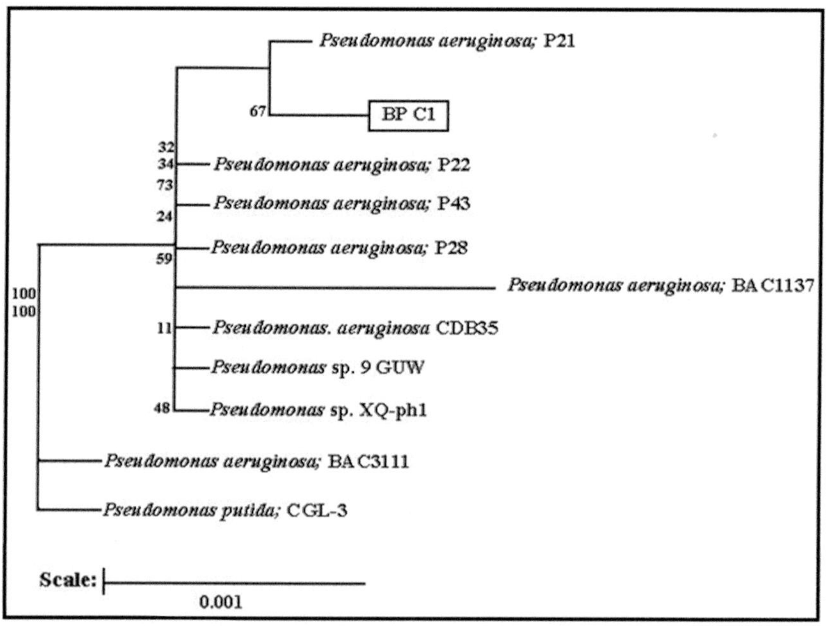

FIGURE 13.3 Phylogenetic relationships of strain BP C1 and other related *Pseudomonas* species based on 16S-rDNA sequencing. The tree was generated using the neighbor-joining method.

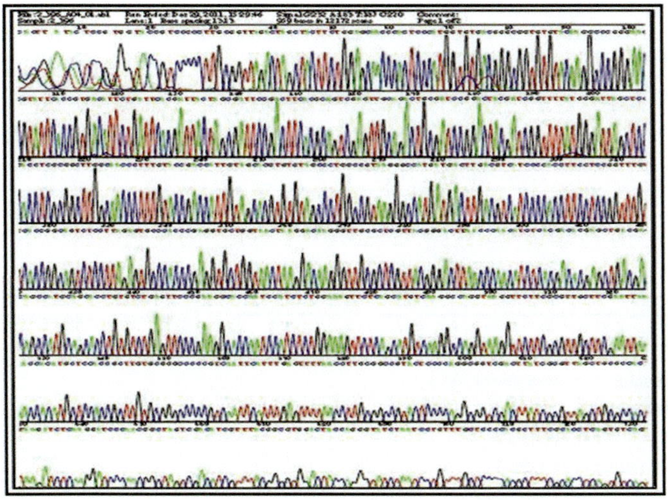

FIGURE 13.4 Electropherogram of 16S-rRNA gene of strain BP C1.

13.4.3 PHYLOGENY OF STRAIN BPC2

A homologous search result of 16S-rDNA sequence of strain BPC2 demonstrated that 98–99% similarity was observed with other species of the genus *Pseudomonas* as shown in Table 13.5. The phylogenetic tree constructed from the 16S-rDNA partial sequence data by the neighbor-joining method. It was observed that *Pseudomonas aeruginosa*; MTH8 (HQ202541) showing 99% homology represented the closest phylogenetic neighbors of the strain BP C2 (Figures 13.5 and 13.6).

TABLE 13.5 Homologous Search Results of 16S-rRNA Gene Partial Sequence of Strain BP C2 Using BLAST from NCBI*

Accession	Organism Name	S_ab Score
DQ237950	*Pseudomonas* sp. Bxl-1	0.999
FJ462714	*P. aeruginosa* Y-4	1.000
GQ926937	*P. aeruginosa* DM2	0.998
GU272400	*Pseudomonas* sp. YKM_M4	0.997
GU726840	*P. aeruginosa* 1242	1.000
HQ202541	*P. aeruginosa* MTH8	0.997
HM448980	*Pseudomonas* sp. NEHU BSSRJ.2	0.999

TABLE 13.5 *(Continued)*

Accession	Organism Name	S_ab Score
HM448981	*Pseudomonas* sp. NEHU BSSRJ.3	0.999
HM448983	*Pseudomonas* sp. NEHU BSSRJ.5	0.999
JF331665	*P. aeruginosa* NL01	0.991

*The 16S-rDNA sequences from microbes showing up to 99% identity is shown.

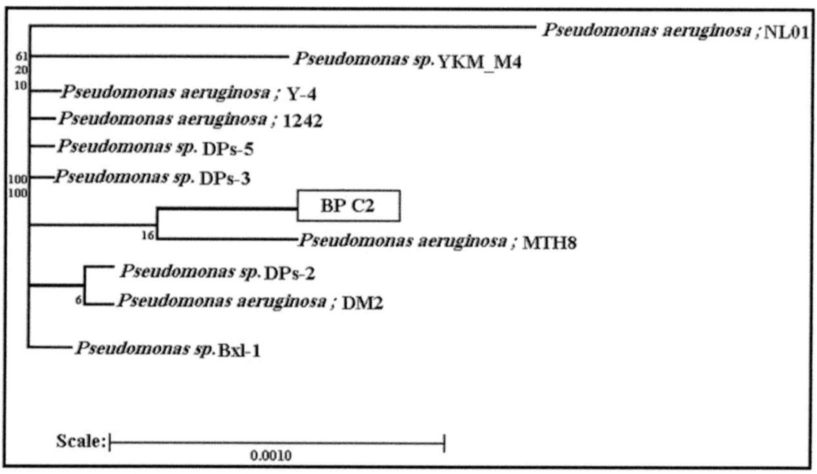

FIGURE 13.5 Phylogenetic relationship (neighbor-joining method) of the strain BP C2 and other closely related *Pseudomonas* species based on 16S-rDNA sequencing.

13.5 BACTERIAL IDENTIFICATION AND RE-DESIGNATION

The 16S-rDNA sequences of *P. aeruginosa* strain BPC1 and *P. aeruginosa* strain BPC2 were deposited to GenBank NCBI and GenBank ID of partial 16S-rDNA sequences are *P. aeruginosa* BPC1 (JQ796859) and *P. aeruginosa* BPC2 (JQ866912) respectively.

13.5.1 TAXONOMIC IDENTIFICATION OF STRAIN BP2

The strain BP2 was previously isolated and taxonomically identified in IMTECH, Chandigarh as *B. circulans* MTCC8167. Further this was screened for biosurfactant production. But biopolymer production from this strain was investigated for the first time.

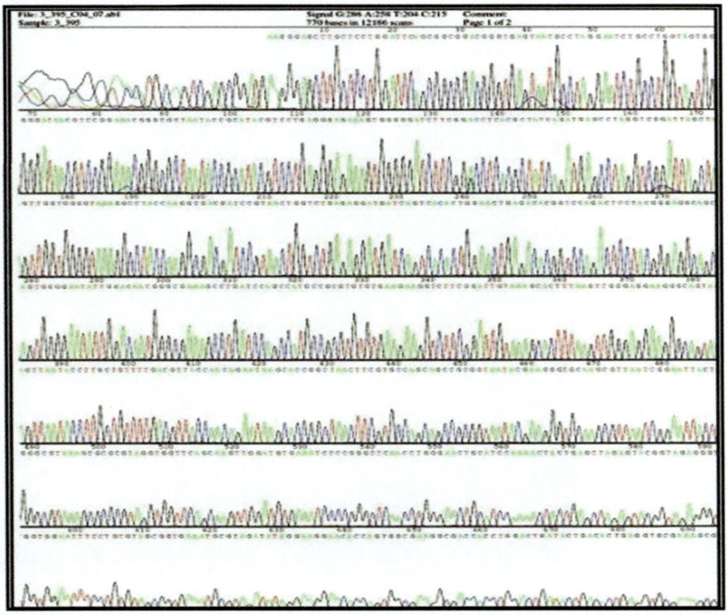

FIGURE 13.6 Electropherogram of 16S-rRNA gene of strain BP C2.

13.5.2 PCR-BASED IDENTIFICATION OF PHA BIOSYNTHETIC GENES IN P. AERUGINOSA STRAINS

PCR was performed using the Forward (I-179L; $5'$-ACAGATCAA-CAAGTTCTACATCTT CGAC-$3'$) and the reverse (I-179R; $5'$-GGTGTT-GTCGTTGTTCCAGTAGAGGATGTC-$3'$) primers based on two highly conserved sequences deduced from a multiple alignment analysis of *Pseudomonas* PhaC1 and PhaC2 genes. The PCR product obtained from the bacterial strains *P. aeruginosa* strains JQ796859 and JQ866912 was about 540 bp (Figure 13.7). The size of the PCR products agrees with the length of the PhaC1 and PhaC2 genes flanked by the I-179L/I-179R primer-pair and evidence for the presence of PhaC genes in the above two *P. aeruginosa* strains.

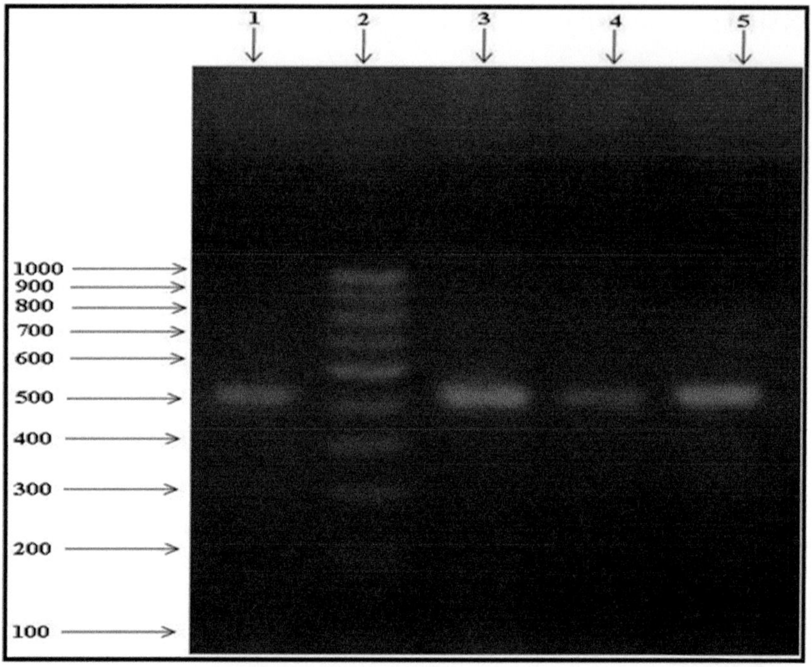

FIGURE 13.7 PCR detection of PhaC1/PhaC2 genes. Purified genomic DNA and regular *Taq* DNA polymerase used in the reactions. Lane 1, 5: *P. aeruginosa* JQ866912; Lane 2: DNA size marker; Lane 3, 4: *P. aeruginosa* JQ796859.

KEYWORDS

- **biopolymer**
- **biosynthetic genes**
- **polyhydroxyalkanoates**
- **pseudomonas**
- ***Pseudomonas aeruginosa***
- **sabouraud dextrose**
- **tris-acetate-EDTA**

REFERENCES

Dubnau, D., et al., (1965). Gene conservation in *Bacillus* species, I. conserved genetic and nucleic acid base sequence homologies. *Proc Natl Acad Sci U.S.A., 54*, 491–497.

Hoffmann, N., et al., (2000). The *Pseudomonas aeruginosa* phaG gene product is involved in the synthesis of polyhydroxyalkanoic acid consisting of medium-chain-length constituents from non-related carbon sources. *FEMS Microbiol. Lett., 184*, 253–259.

Langenbach, S., et al., (1997). Functional expression of the PHA synthase gene phaC1 from *Pseudomonas aeruginosa* in *Escherichia coli* results in poly (3-hydroxyalkanoate) synthesis. *FEMS Microbiol. Lett., 150*, 303–309.

McCool, G. J., & Cannon, M. C., (2001). PhaC and PhaR are required for polyhydroxyalkanoic acid synthase activity in *Bacillus megaterium. J. Bacteriol., 183*, 4235–4243.

Peoples, O. P., & Sinskey, A. J., (1989). Poly-beta-hydroxybutyrate (PHB) biosynthesis in *Alcaligenes eutrophus* H16. Identification and characterization of the PHB polymerase gene (phbC). *J. Biol. Chem., 264*, 15298–15303.

Pohlmann, A., et al., (2006). Genome sequence of the bioplastic-producing "Knallgas" bacterium *Ralstonia eutropha* H16. *Nat. Biotechnol., 24*, 1257–1262.

Qi, Q., & Rehm, B. H. A., (2001). Polyhydroxybutyrate biosynthesis in *Caulobacter crescentus*: Molecular characterization of the polyhydroxybutyrate synthase. *Microbiology, 147*, 3353–3358.

Qi, Q., et al., (1997). Synthesis of poly (3-hydroxyalkanoates) in *Escherichia coli* expressing the PHA synthase gene phaC2 from *Pseudomonas aeruginosa*: Comparison of PhaC1 and PhaC2. *FEMS Microbiol. Lett., 157*, 155–162.

Sambrook, et al., (2001). *Molecular Cloning: A Laboratory Manual.* Cold Spring Harbor Laboratory Press.

Schubert, P., et al., (1988). Cloning of the *Alcaligenes eutrophus* genes for synthesis of poly-beta-hydroxybutyric acid (PHB) and synthesis of PHB in *Escherichia coli. J. Bacteriol., 170*, 5837–5847.

Schwartz, E., et al., (2003). Complete nucleotide sequence of pHG1: A *Ralstonia eutropha* H16 megaplasmid encoding key enzymes of H (2)-based ithoautotrophy and anaerobiosis. *J. Mol. Biol., 332*, 369–383.

Slater, S. C., et al., (1988). Cloning and expression in *Escherichia coli* of the *Alcaligenes eutrophus* H16 poly-beta-hydroxybutyrate biosynthetic pathway. *J. Bacteriol., 170*, 4431–4436.

Vaneechoutte, M., et al., (2004). Wautersia gen. nov., a novel genus accommodating the phylogenetic lineage including *Ralstonia eutropha* and related species, and proposal of *Ralstonia* [*Pseudomonas*] syzygii comb. Nov. *Int. J. Syst. Evol. Microbio., 54*, 317–327.

Wang, Y., et al., (2006). Construction of recombinant *Bacillus subtilis* for production of polyhydroxy- alkanoates. *Appl. Biochem. Biotechnol., 129–132*, 1015–1022.

CHAPTER 14

PHAs in Enhancing the Stability of Colloidal Silver Nanoparticles (SNPs)

The PHA suspension (0.5%) in distilled water was prepared by sonication for 3 h, using ultrasonic water bath homogenizer). The PHA inclusion bodies in the bacterial species *B. circulans* (MTCC 816, stained with Nile blue A and observed under the fluorescent light (100X magnifications). From the PHA stock, an aliquot of 5 ml was made up to 30 mL to obtain 0.002 M sodium borohydride ($NaBH_4$). The flask was placed into an ice bath and cooled for 20 min. The assembly was stirred gently using a magnetic stirrer. Now, 10 mL of 0.001 M $AgNO_3$ solutions was added drop wise, about 1 drop per second, until the whole amount was used up. After the addition of the silver nitrate, the solution turned light yellow in color and the SNP was synthesized in the PHA dispersed colloidal solution.

14.1 FOURIER-TRANSFORM-INFRARED (FTIR) SPECTROSCOPIC ANALYSIS

The FTIR spectra in Nicolett Impact 410 spectrometer of the biopolymer and SNPPHA samples were recorded using KBr pellet. Spectra showing the functional groups were assessed to determine the composition of biopolymer. The spectra were obtained in wave no (cm^{-1}) 400 to 4,000. Dispersed the solid sample was distributed uniformly in the dry matrix KBr nujol (KBr) mull, and compressed to result transparent disc.

Bacterial Biopolymers. B. K. Konwar, PhD (Author)

14.2 STABILITY ANALYSIS OF SNP SOLUTION USING UV-VIS SPECTROPHOTOMETER

Three different solutions of SNP were prepared in triplicate, each containing 0.08% PHA in 30 mL distilled water. The bright yellow color of the solution is a proof of the presence of SNP with a maximum absorption at about 400 nm, and the solution was labeled as SNPPHA in situ. Another solution of SNP synthesized by mixing pure SNP colloids and PHA-dispersed solution of 0.25% (w/v) at 1:1 was labeled as SNPPHA mixture. The analysis was carried out with the absorption maxima shifted towards longer WL and simultaneously the color of the colloid changed to violet and sometimes grayish as the particles settled down. Based on the spectroscopic data, all three types of SNP colloidal solutions (SNP, SNPPHA mixture and SNPPHA in situ) were recorded for 30 days with an interval of 5 days in between.

14.3 TEM ANALYSIS

The colloid of PHA-dispersed solution was spread on TEM (transmission electron microscope). The stub and sample preparations were done using the standard protocol. Consequently, each sample was examined.

14.4 PHOTOLUMINESCENCE (PL) STUDY OF THE NANOPARTICLES (NPS)-BIOPOLYMER HYBRID

The fluorescence behavior of PHA polymer extracted from *P. aeruginosa* JQ866912 was studied for the development of a method to estimate different nanoparticles (NPs) in solution. PHA polymer was prepared by dissolving in chloroform (1.4%). The metal oxide NPs ZnO, NiO, and CuO were used. The different concentrations of ZnO, NiO, and CuO were mixed with the polymer solution by constant sonication for about 30 min. All PHA-nanoparticle hybrid mixtures were cast on a glass slide and dried in vacuum for the further characterization.

14.5 CHARACTERIZATION OF THE HYBRID

For the determination of the size of the nanomaterials, XRD was used. By this technique, the crystal size of the nanomaterials and average size of the particles in the matrix can be detected. XRD study for the biopolymer-NPs was done using Miniflex, Rigaku (Japan) with CuK radiation (= 0.15418 nm) at 30 kV and 15 mA with scanning rate of 0.005 S^{-1} in a 2 range of 3,070°.

The optical properties of the PHA-NPs hybrid materials are determined by UV-vis spectroscopy and PL spectroscopy. The distribution of the NPs within the PHA biopolymer matrix and the surface study of the hybrid materials are analyzed by using SEM.

14.6 PHA IN COMPOSITE-BASED BIOSENSOR PREPARATION WITH GOLD NANOPARTICLES (AUNPS)

Artemisinin (99% purity) HRP, aniline monomer (99.5% purity), ITO coated glass plates were used (Sigma-Aldrich, Germany). Serum sample was obtained from the healthy volunteers and kept frozen at–20°C. The serum sample after gentle thawing was spiked with required concentrations of the drug. Stock solution of artemisinin was prepared by direct dissolution of 1 mg.ml^{-1} of artemisinin in fresh 0.05 M PO_4 buffer (PBS) having pH 7.0 ± 0.1. The standard working solutions were prepared with appropriate dilutions of the stock. Potassium ferrocyanide (Fe(CN)63-/64-)-containing PBS of 200 ml with 0.05 M ionic strength in the pH range 5.5–9 were prepared in deionized water (TKA Millipore) by adding measured amounts of NaH_2PO_4, Na_2HPO_4 and adjusted with 1M HCl and 0.1 M NaOH.

14.7 CHARACTERIZATION OF AUNP BIOSENSOR

Using the Autolab Potentiostat/Galvano-stat electrochemical measurements, the autochemical (a) NOVA software (Eco Chemie, Netherlands) and (b) Gamry Echem Analyst Software (Ref 3000, USA) wherein the working electrode was tested with PHA/ITO, PHA/AuNPs/ITO and PHA/AuNPs/HRP/ITO. A platinum wire and Ag/AgCl (3 M KCl) were the counter and reference electrodes, respectively Dr Bob's electrochemical cell stand 20 ml was used to carry out CV and EIS. Surface topology of films was studied using a SEM (JEOL-JSM-6390LV).

14.8 PREPARATION OF PHA-AU NANOCOMPOSITE FILM

The *P. aeruginosa* strain JQ866912 PHA solution 20 μl (3 mg.ml^{-1}) and 80 μl of AuNPs in water (3 mg ml^{-1}) were mixed and sonicated for 30 min. The mixture was cast over 0.25 cm^2 two ITO glass plates with normal casting and two deep casted in a special chamber.

14.9 IMMOBILIZATION OF HRP ON PHA-AU NANOCOMPOSITE FILM

The bio-electrode was prepared by adsorption technique (in a specially assembled cell) immobilizing HRP enzyme 1 mg.ml^{-1} solution in PO_4 buffer (pH 7.0 ± 0.1) onto PHA-Au nanocomposite, incubated overnight at 4°C. In one experiment cross-linker was glutaraldehyde, in the other none. The condition for enzyme immobilization was used based on prior studies (Lin et al., 2007; Zhang et al., 2005). To remove the loosely bound materials, the biosensors were rinsed with PO_4 buffer and preserved at 4°C with pH 7 ± 0.1 for further use.

14.10 PREPARATION OF ARTEMISININ SOLUTION WITH PHA-AU NANOCOMPOSITE

The solution of artemisinin in water (0.0011 mg/100 ml) was prepared by sonicating for 3 h. The stock artemisinin solution was added with freshly prepared 0.05 M PBS (pH 7 ± 0.1) to make the required concentrations. Human serum samples were centrifuged, and added 10 times buffer solution prior to measurements. The standard method was used to analyze pharmaceutical samples and artemisinin-spiked human serum samples for validating the modified electrode.

14.11 DEGRADATION OF COMPLEX SYNTHETIC POLYMERS

The growth of *P. aeruginosa* strain OBP1 on various nanocomposite films such as hyperbranched epoxy (HBE), modified hyperbranched epoxy (MHBE) and their clay nanocomposites along with the corresponding pristine polymeric films can be realized. Growth curves of bacterial strain

OBP 1 on HBE, MHBE 30, NCE 2.5 and NCME 2.5 their nanocomposites. Values are the mean of three independent experiments ± standard deviation.

The bacterial strain on the tested polymers increased with the increase in exposure time to the bacterial strain. The growth rate in all the tested nanocomposites as well as on the pristine polymers was not significant up to two weeks of bacterial exposure. But after 2–3 weeks of treatment, the biodegradation rate increased sharply as could be realized from the bacterial population density determination by McFarland turbidity method. The biodegradation process caused severe damage to almost all nanocomposites as revealed by SEM of the recovered nanocomposite films following five weeks of bacterial attack. The same was further supported by the decrease in the weight of the nanocomposite films after biodegradation. However, the extent of biodegradation was found to be low in the case of pristine hyperbranched epoxy/clay nanocomposite in *P. aeruginosa* was reported to be a versatile bacterial strain capable of utilizing various types of carbon sources ranging from simple glucose to complex petroleum-based hydrocarbons (Jendrossek and Handrick, 2002; Bordoloi and Konwar, 2008; Bharali and Konwar, 2011). From biodegradation studies of synthetic MHBE/OMMT clay nanocomposite-based polymers, the bacterial isolate *P. aeruginosa* strain OBP1 can utilize the synthesized polymers as is evident with the increase in population density. The rate of bacterial cell density increases with the duration of the treatment time. This is particularly observed in the case of MHBE nanocomposites due to the catalytic role of clay in the hydrolysis of the ester groups present in the modified systems (Roy et al., 2013). The presence of terminal hydroxyl groups in the clay layers can cause heterogeneous hydrolysis after absorbing water in the presence of microbes. This process is known to require some time for the initiation, which might be the reason for the remarkable improvement of biodegradation after 2–3 weeks of bacterial treatment (Philip et al., 2007; Kim and Rhee, 2003). The extent of biodegradation was comparatively low in the case of pristine hyperbranched epoxy/clay nanocomposite in comparison to the MHBE nanocomposites. This was mainly due to the presence of polar ester groups and fatty acid chain moieties mainly caused this in vegetable oil-based polyester MHBE/clay nanocomposites, which attract microbes.

14.12 PHA OF *B. CIRCULANS* IN STABILITY ENHANCEMENT OF COLLOIDAL SNP

14.12.1 CHARACTERIZATION BY USING FTIR SPECTROSCOPY

The FT-IR spectrum of SNP confirmed the presence of silver in the composite. The absorption picks confirmed the presence of silver in SNP, SNP-PHA in situ and SNP-PHA mix (Figure 14.1). The shifting and decrease in the intensity of absorption peaks revealed the successful conjugation or the presence of SNP in the SNP-PHA colloids. The photographic evidence (Figure 14.2) depicted the stabilization of SNP colloids for 30 days.

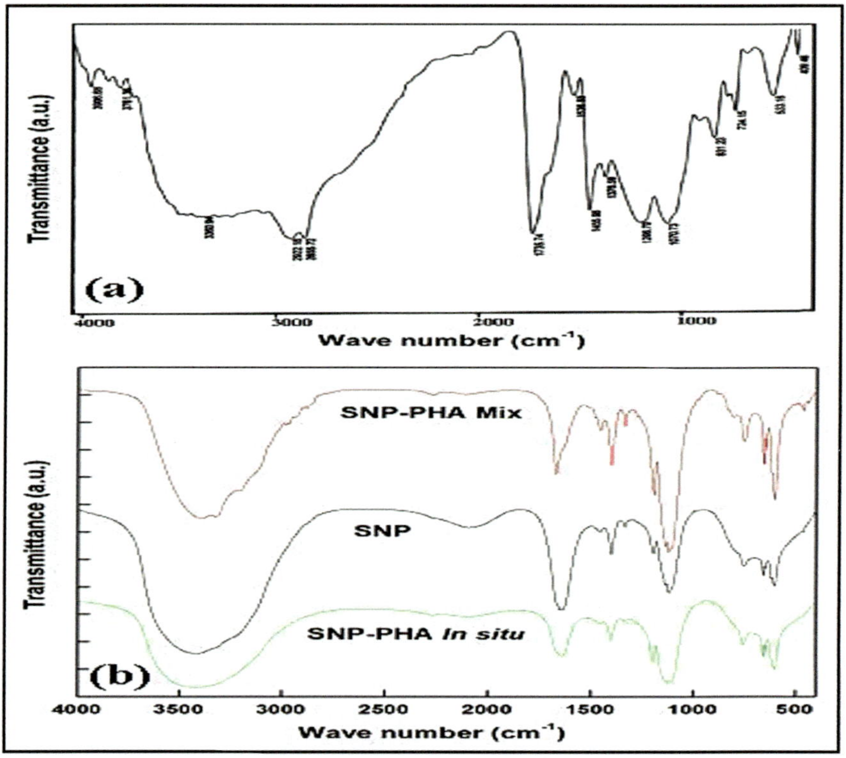

FIGURE 14.1 FT-IR spectrum of PHA polymer isolated from *B. circulans* MTCC8167 (a); and the SNP, SNP-PHA *in situ* and SNP-PHA mixture (b).

FIGURE 14.2 SNP in different colloids: (a) SNP-PHA *in situ*; (b) SNP-PHA mixed; and (c) SNP after 30 days aging at room temperature.

14.12.2 CHARACTERIZATION OF SNP-PHA BY TEM

The size and morphology of the SNP are analyzed by TEM and data thus obtained are presented in Figure 14.3. TEM image confirms the presence of SNP in SNP-PHA in situ colloid.

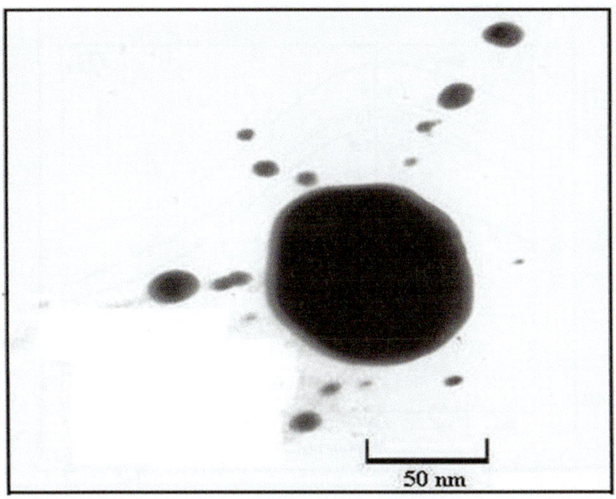

FIGURE 14.3 Transmission electron micrograph (TEM) of the silver nanoparticle in SNP-PHA *in situ*.

14.12.3 CHARACTERIZATION OF SNP-PHA BY UV-VIS SPECTROSCOPY

The UV-Vis spectra of the SNP colloids are presented in Figure 47(A). After 29 days the spectrum shifted towards 420 nm. It could be due to aggregation of SNP in the colloid. In SNP-PHA in situ, the spectra showed the maximum wavelength near 400 nm in day 1 consistent up to 30 days (Figure 14.4(B)).

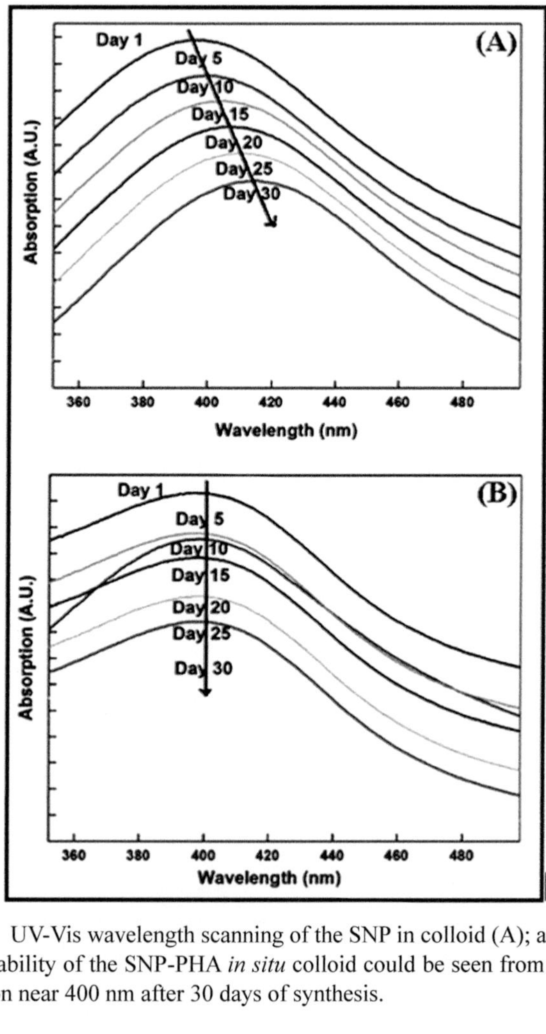

FIGURE 14.4 UV-Vis wavelength scanning of the SNP in colloid (A); and SNP-PHA *in situ* (B). The stability of the SNP-PHA *in situ* colloid could be seen from the consistency of the absorption near 400 nm after 30 days of synthesis.

14.13 PHOTOLUMINESCENCE (PL) OF *P. AERUGINOSA* JQ866912 PHA-BASED NANOPARTICLES (NPS)

14.13.1 XRD ANALYSIS OF THE PHA-METAL OXIDE NANOPARTICLE HYBRIDS

The crystalline nature of the PHA/NPs hybrids, the thin film on the glass substrate was used for the XRD analysis. The XRD pattern of the prepared virgin PHA biopolymer, virgin NPs and PHA/NiO NPs, PHA/ZnO NPs and PHA/CuO NPs hybrids with different concentrations is shown in Figure 14.5. The comparative study with the virgin NPs, virgin polymer sample and the hybrid materials show the successful incorporation of the NPs in the polymer matrix, indicating that the crystal planes arise from the process.

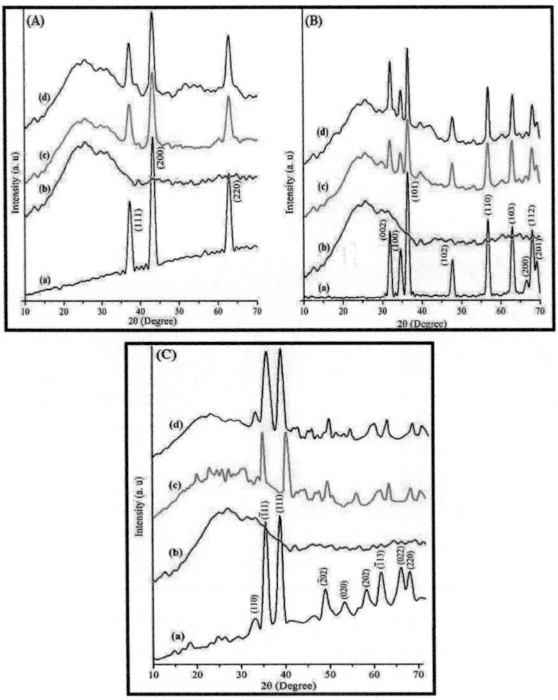

FIGURE 14.5 (A) XRD for NiO NPs (a), PHA (b), PHA/NiO NPs (1:1) hybrid (c), PHA/NiO NPs (1:3) hybrid (d); (B) XRD for ZnO NPs (a), PHA (b), PHA/ZnO NPs (1:1) hybrids (c), PHA/ZnONPs (1:3) hybrid (d); and (C) XRD for CuO NPs (a), PHA (b), PHA/CuO NPs hybrid (1:1) (c), PHA/CuO NPs hybrid (1:3) (d) materials.

14.13.2 OPTICAL PROPERTIES OF THE PHA-METAL OXIDE NANOPARTICLE HYBRIDS

The absorption and emission spectra of the polymer-metal nanoparticle hybrids were recorded with UV-vis spectrophotometer and Fluorescence spectrophotometer. The UV absorbance spectra (Figure 14.6) showed shifting of the absorption maxima of the PHA polymer when incorporated with the metal NPs. The PL spectra (Figure 14.7) of the nano-polymer hybrids were recorded by preparing thin films over glass plates. The PL spectra of the material showed a dramatic change in the emission nature to that of the virgin polymer. The intensity of the PL spectra increases along with an increase in the concentration of NPs.

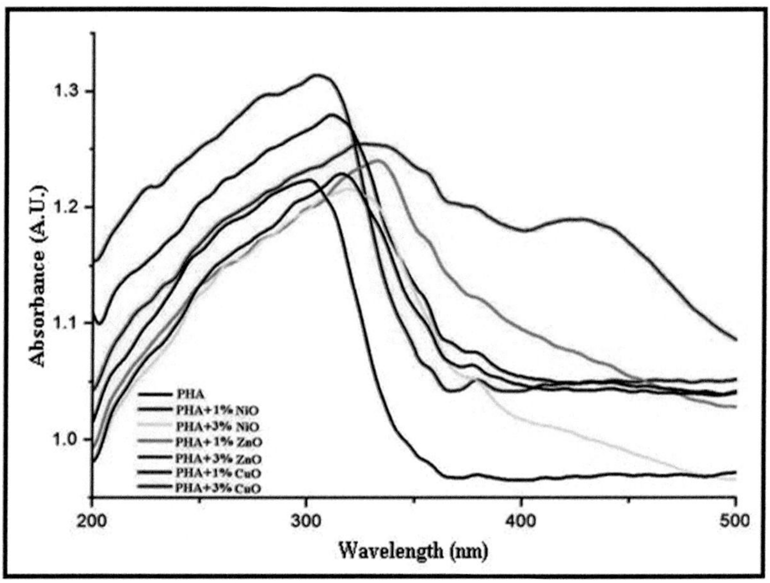

FIGURE 14.6 UV-vis spectra of crude polymer and polymer with different concentration of different nanoparticles in chloroform.

The SEM micrograph of the pure polymer and polymer-nanoparticle hybrids is presented in Figure 14.8. The particle sizes originated in the image demonstrated the presence of NPs in the matrix.

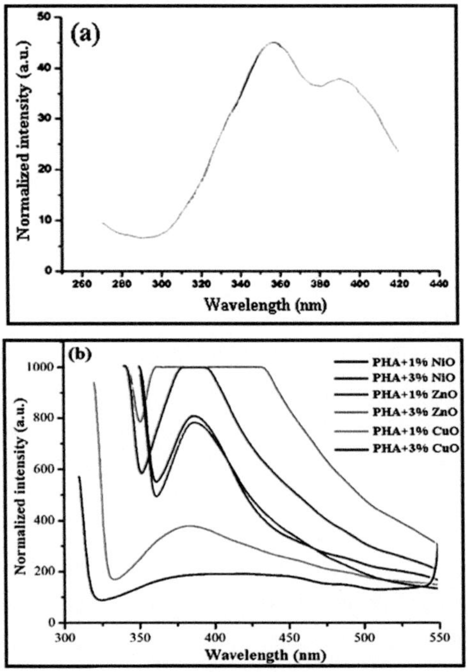

FIGURE 14.7 Fluorescence spectra of (a) polymer; (b) polymer in the presence of various nanoparticles in chloroform at their excitation wavelength SEM analysis.

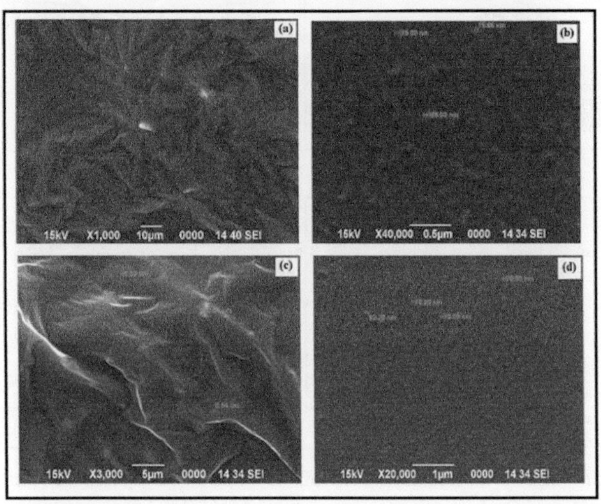

FIGURE 14.8 SEM photograph of crude PHA (a); PHA/NiO NPs (b); PHA/ZnO NPs (c); and PHA/CuO NPs (d) hybrids.

14.14 PHA-BASED AUNP BIOSENSOR IN THE DETECTION OF ANTIMALARIAL DRUG-ARTEMISININ

14.14.1 CYCLIC VOLTAMETRIC (CV)

The electrochemical behavior of the PHA/AuNPs/HRP/ITO based biosensor was investigated by using CV. The CV behavior of (A) PHA/ITO, (B) PHA/AuNPs/ITO and (C) PHA/AuNPs/HRP/ITO films was studied with variable scan rates of 5 to 40 mV s^{-1} in phosphate buffer solution (PBS) [50 mM, pH 7.0 ± 0.1 and, 0.9% NaCl] having 5 mM of 5 mM $[Fe(CN)_6]^{3-/4-}$ (Figure 14.9). The shift of the anodic peak potential (Figure 14.9(A–C)) was observed towards the positive side as well as the cathodic peak potential shifts in the reverse direction. The pH effect on the peak current was assessed in the range of 5.5–9 in the presence of 0.01 μg mL^{-1} artemisinin (Figure 14.9(D)). Plotting of current vs pH showed the presence of the maximum peak current and the good pH 7.0 ± 0.1 peak shape of PBS. The redox peak currents could be proportional to the square root of scan rate, 1/2 (Figure 14.10) indicating a diffusion electron transfer process.

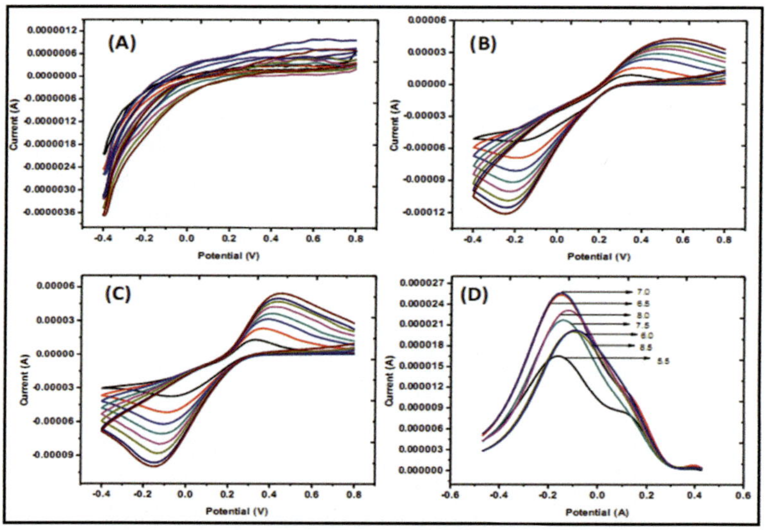

FIGURE 14.9 Cyclic voltammograms of (A) PHA/ITO; (B) PHA/AuNPs/ITO; (C) PHA/AuNPs/HRP/ITO electrodes in PBS (50 mM, pH 7.0 ± 0.1) containing (5 mM) $[Fe(CN)_6]^{3-/4-}$ at scan rate from 5 mV s^{-1} to 40 mV s^{-1}; and (D) study of pH.

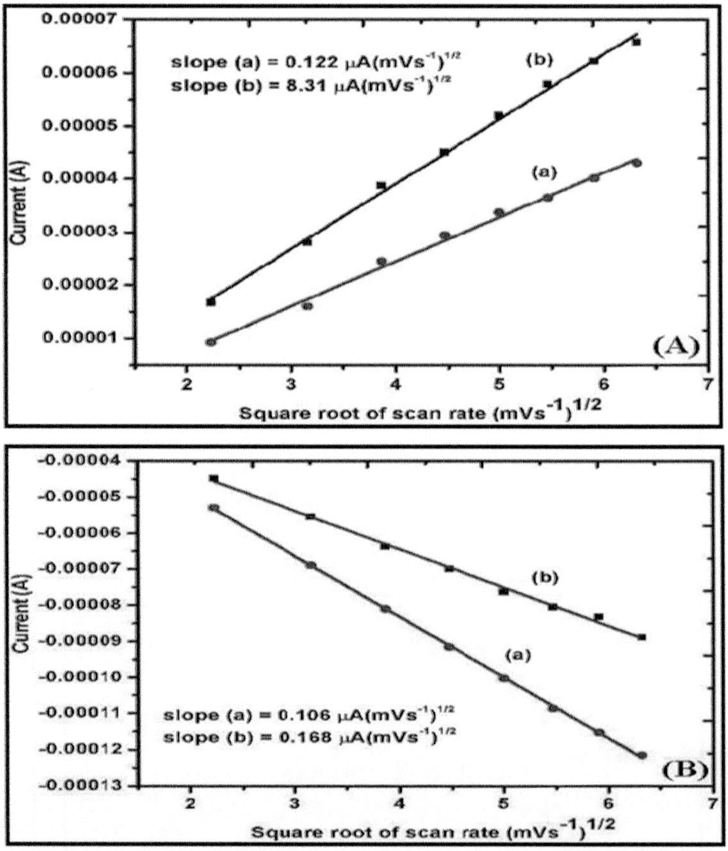

FIGURE 14.10 Oxidations (A); and reduction (B) peaks current with square root of scan rate for PHA/AuNPs/ITO (a); and PHA/AuNPs/HRP/ITO (b).

14.14.2 EIS

The effect of surface charge modulation was investigated to determine the relative variation in CV at constant scan rate (5 mVs^{-1}) in PBS buffer containing 5 mM [Fe $(CN)_6]^{3-/4-}$ for electrodes (A) PHA/ITO, (B) PHA-AuNPs/ITO and (C) PHA-AuNPs-HRP/ITO (Figure 14.11(A)) which also showed enhancement of current in PHA-AuNPs/ITO electrode in comparison to the others. The determination of relative change in the surface resistance at zero potential EIS was carried out on (a) PHA/ITO, (b) PHA/AuNPs/ITO and (c) PHA/AuNPs/HRP/ITO films (Figure 14.11(B)).

From the figure, the surface charge transfer resistance, RCT in the impedance spectrum of PHA/AuNPs/ITO electrode is observed to be lower in comparison to the others which depicted the successful incorporation of AuNPs into the PHA films.

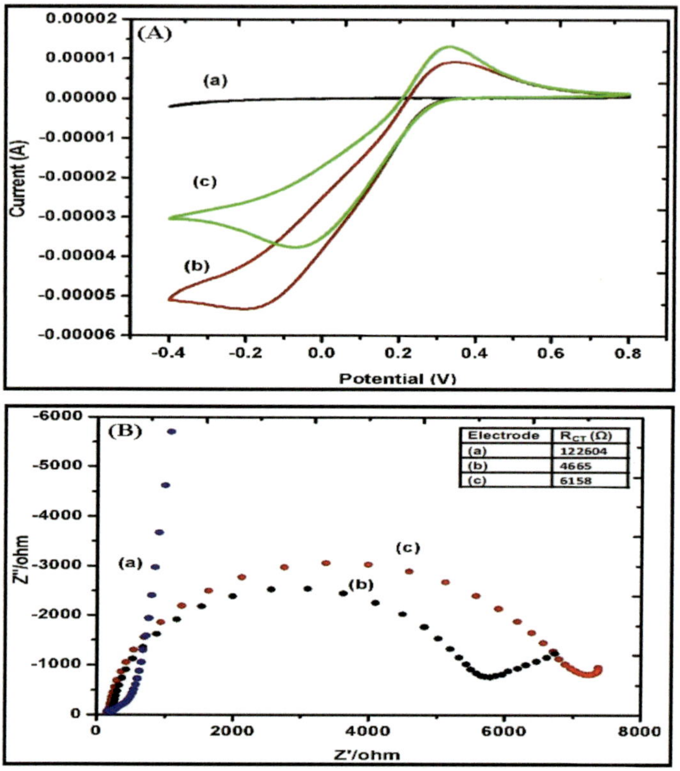

FIGURE 14.11 Cyclic voltammograms of (A): (a) PHA/ITO; (b) AuNPs/PHA/ITO; (c) AuNPs/PHA/HRP/ITO scan rate at 5 mV s^{-1}; and (B) electro-chemical Nyquist plots of (a) PHA/ITO; (b) AuNPs/PHA/ITO; (c) AuNPs/PHA/HRP/ITO electrodes in PBS buffer (50 mM, pH 7 ± 0.1) containing 5 mM $[Fe(CN)_6]^{3-/4-}$.

14.14.3 SEM

The surface morphology of the working electrodes was studied with the SEM micrograph. Figure 14.12(a) shows non-porous and smooth surface of the PHA, Figure 14.12(b) the surface of electrodes consisting of uneven porous matrix of PHA, embedded with AuNPs and Figure 14.12(c) immobilization of HRP to PHA/AuNPs film.

14.14.4 APPLICATION TO REAL SAMPLE

The applicability of electro-chemical biosensor based on PHA/AuNPs/HRP/ITO bio-electrode was assessed for antimalarial drug artemisinin in the bulked and spiked human serum. Voltammograms and calibration curves are presented in Figure 14.13(a) and (b). The figures reveal current being proportional to artemisinin concentration over a convenient range. With the calibration curve and following the standard addition method, the precision was estimated for 0.01–0.08 μg mL^{-1} of the drug. The calibration equation parameters and the related validation data are presented in Table 14.1.

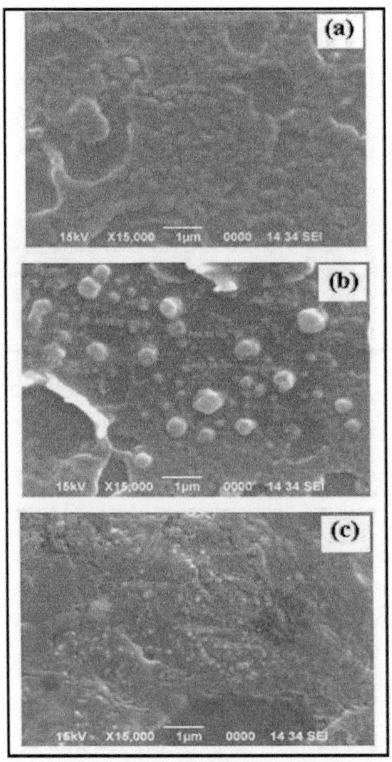

FIGURE 14.12 SEM micrograph of the surface of (a) PHA/ITO; (b) PHA/AuNPs/ITO; (c) PHA/AuNPs/HRP/ITO biofilms.

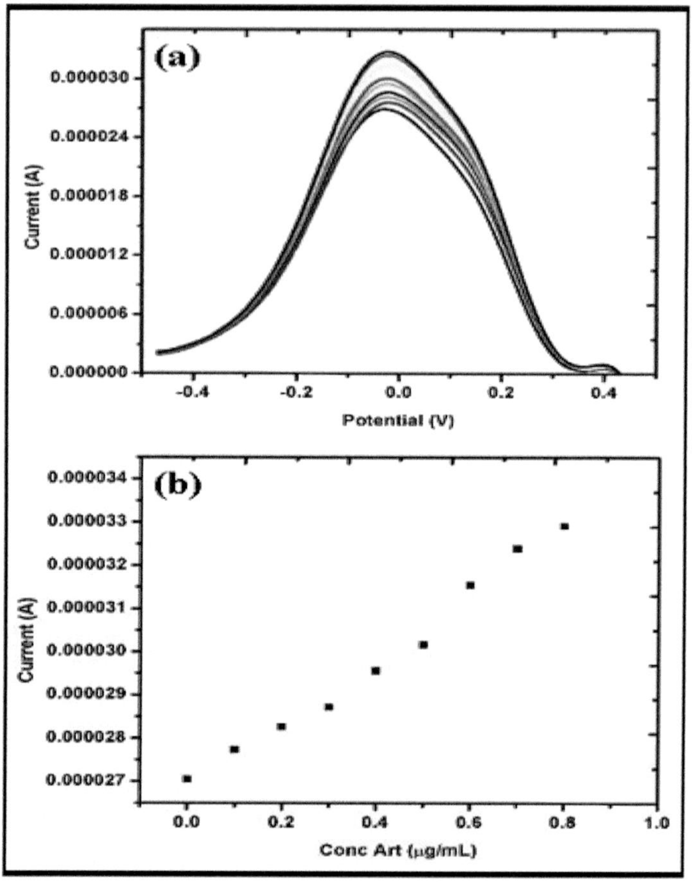

FIGURE 14.13 (a) Biosensors response with increase in concentration of artemisinin; and (b) calibration curve shows linearity for artemisinin as 0.01 to 0.08 μg mL^{-1}.

TABLE 14.1 Regression Data of the Calibration Lines for the Quantitative Determination of Artemisinin by Organic-Inorganic Hybrid Nanocomposite AuNPs/PHA/HRP-Based Biosensor in Bulk Form and Spiked Human Serum Using DPV Waveform

Parameters	Bulk Form	Human Serum
Linearity range (μg mL^{-1})	0.01–0.08	0.01–0.06
Slope(A/μgmL^{-1})	8.8641×10^{-6}	9.1050×10^{-6}
Intercept (A)	2.6222×10^{-5}	2.6172×10^{-5}

TABLE 14.1 *(Continued)*

Parameters	Bulk Form	Human Serum
Correlation coefficient (r^2)	0.985	0.984
RSD of the intercept	1.0438×10^{-8}	1.0762×10^{-8}
RSD of the slope	3.8728×10^{-7}	4.6491×10^{-7}
Reproducibility (%RSD)	1.01	1.13
Repeatability (%RSD)	0.66	1.01
LOD (μg mL^{-1})	0.0035	0.0036
LOQ (μg mL^{-1})	0.0110	0.0121

14.15 PHA IN INCREASING STABILITY OF SNP

14.15.1 CHARACTERIZATION BY USING FTIR SPECTROSCOPY

The presence of silver in the SNP was confirmed by FTIR spectrum. The absorption picks near 600 cm^{-1} indicating NH stretching, 1,100 cm^{-1} assigned as the absorption peaks of CO; on the other hand, the stretching vibration of C = C was indicated by peaks of 1,400 and 1,600 cm^{-1}. Therefore, this absorption peaks revealed the presence of silver in the SNP and SNP-PHA colloids (Figure 14.7). The peak near 1,666 cm^{-1} of the SNP-PHA mix might be due to the ester carbonyl C = O of the PHA which shifted due to conjugation or presence of SNP. The small valley in SNP was highlighted by the same peak towards smaller wave no. The double dilution of the SNP colloids with PHA dispersed solution could be cause for the low intensity of the SNP peak. The same peak in *in situ* SNP-PHA broadened due to the presence of PHA and SNP. In the case of SNP, the peak near 1,666 cm^{-1} was broad but different from that of SNP-PHA in situ. The pattern clearly revealed the presence of the silver nanoparticles (SNPs) in the medium.

The SNP colloids were stabilized for 29 days by different methods, after the days of incubation, the SNP-PHA in situ remained stable, whereas the other two samples showed color pattern aggregation of the SNP. The bright yellow color of the SNP-PHA in situ suggests the stability of the SNP in the colloid.

14.15.2 CHARACTERIZATION OF SNP-PHA BY TEM

TEM image revealed the existence of SNP in SNP-PHA in situ colloid. NPs were spherical and their size to be 5–60 nm. However, the number of small particles (5–15 nm) was more as compared to larger (60 nm) particles. Saha et al. (2010) reported that the TEM image of SNPs produced by a phytopathogenic fungus *Bipolaris nodulosa* possessed spherical, semi-pentagonal structure having 10–60 nm in diameter. The use of sodium borohydride in SNP as the reducing agent could produce spherical, semi-pentagonal structure (Roy, 2013). They reported that the NPs formed are spherical in shape and uniform having a size range from 7 nm to 25 nm.

14.15.3 CHARACTERIZATION OF SNP-PHA BY UV-VIS SPECTROSCOPY

The UV-V is spectra of SNP colloids showed absorption spectra near 400 nm in day 1 due to its surface plasmon resonance phenomenon indicating particle size. After 29 days, the spectrum shifted slightly towards 420 nm along with a decrease in the surface plasmon absorption intensity. This could be due to the aggregation of SNP in the colloid with time; larger particles with lesser energy and hence longer WL. The aggregation of SNP was stimulated by two different phenomena; (i) reaction of excess sodium borohydride with SNP increasing overall ionic strength (Solomon et al., 2007) and (ii) addition of the electrolytes such as NaCl. Salt shields charges causing the particles to clump together resulting in aggregates. The colloidal solution of silver turned dark yellow, violet, and then grayish whereas in the case of SNP-PHA *in situ*, the UV-Vis spectra showed maximum wavelength near 400 nm in day 1 consistent up to day 30. The color consistency and the range of maximum WL of the SNP-PHA *in situ* indicated their stability throughout the period. Udapudi et al. (2012) reported a lack of change in 30 days peak position of SNPs synthesized with the reductant trisodium citrate containing aqueous solution of silver nitrate which depicted its stability. The presence of hydrophobic PHA in the colloid makes the SNP-PHA highly stable. In the colloids PHA probably acts as a stabilizer providing the surface for immobilization of SNP. The FTIR spectroscopy confirmed it, but there was no formation of new bond, only physical immobilization of SNP in PHA; in *in situ* SNP-PHA, composite mix was evident on filtration through 0.45 μm nitrocellulose membrane. The SNP-PHA mixture after 29 days of aging showed an intermediate absorption maximum shift near 415 nm. The low immobilization of SNP might be the cause for the low stability of the mixed sample in comparison to the *in situ* one. However, the observations need further investigation.

14.16 PHOTOLUMINESCENCE (PL) STUDY OF THE POLYMER ALONG WITH DIFFERENT NANOPARTICLES (NPS)

14.16.1 XRD ANALYSIS OF PHA-METAL OXIDE NANOPARTICLES (NPS) HYBRIDS

To determine the crystalline nature of the PHA/NPs hybrids, the thin film on the glass substrate was used for the XRD analysis. From this study, the diffraction peak at 2 = 38° for the plane, 2 = 43.15° for the plane and 2 at 64.8° for the plane of NiO NPs are observed in case of virgin NPs. But in the case of PHA/NiO hybrid these peaks showed slight shifting and slight decreasing of the peak intensity. 2 = 31.94° (84), 34.55° (87), 36.45° (85), 47.65° (87), 56.75° (94), 62.95°, 66.45° (88), 68.10° (96) and 69.20° (176) of the ZnO confirms the formation of the NPs which were also present in the PHA/ZnO hybrids but slightly at different positions and decrease in their peak intensity. Peaks were observed at 33.33°, 35.69°, 38.90°, 48.91°, 53.36°, 58.24°, 61.72°, 65.8° and 66.2° corresponding to different planes of CuO NPs. This confirms the formation of CuO NPs in the reaction mix. The broadening and shifting of the peaks in the case of PHA-nanoparticle hybrids and the position of the crystal planes prove the formation of the nanoparticles within the polymer.

14.16.2 OPTICAL PROPERTIES OF PHA-METAL OXIDE NANOPARTICLE HYBRIDS

From the UV absorbance spectra, it was revealed that the absorption maximum of the PHA polymer has been shifted when incorporated with the metal NPs. The polymer absorbed in the region of 285 nm and 295 nm. The absorption intensity of the polymer became the highest with the incorporation of 1% NiO NPs with the polymer than the other two NPs. But there was a decrease in the intensity of the polymer hybrid with the incorporation with 3% NiO NPs. The increased concentration of NiO NPs affect the UV absorption as NiO NPs do not absorb in the UV region. Therefore, it decreases the absorption intensity of the polymer/metal oxide NPs hybrid. The other two polymer hybrid NPs (CuO and ZnO) possessed the same effect as that of the NiO hybrid material, but their absorption spectra showed shifting of the absorption maxima as compared to the crude polymer material.

The PL spectra of the material showed a dramatic change in the emission nature to that of the virgin polymer which accounts for the interaction of the metal nanoparticle with the polymer. The change in the surface states of the metal NPs enhanced the luminescence of the polymer-metal oxide NPs hybrids. The density of the surface would have increased with the decrease in the particle size due to the large surface to volume ratio (Green, 1995). The PL spectra clearly exhibited with the increase in NPs concentration, the intensity of PL also increased. This was due to energy transfer mechanism. The polymer materials generally form clusters and transfer their energy to each other and thereby causing a decrease in PL intensity. But when NPs incorporated with the polymer, NPs break the cluster and allowing polymer chains to absorb and emit light independently. Thereby, the enhancement of PL intensity takes place.

14.16.3 SEM ANALYSIS

The surface morphology of the polymer-metal oxide NPs hybrids thin film was investigated using SEM. The SEM micrograph shows the presence of NPs within the polymer matrix. The SEM images also show the well distribution of the NPs within the polymer matrix. The size of the particles was found to be about 70–92 nm. Kausik et al. (2008) have studied the surface morphology of polyaniline-ZnO nanocomposite and suggested that embedding of NPs in the polymer matrix could impart novel properties important for sensing application.

14.17 BIOSENSOR WITH GOLD NANOPARTICLES (AuNPS)

14.17.1 CYCLIC VOLTAMMETRIC (CV) STUDY

The CV electrochemical behavior of (a) PHA/ITO (b) PHA/AuNPs/ITO and (c) PHA/AuNPs/HRP/ITO electrode was studied at the scan rate 5–40 mV s^1 in PBS (50 mM, pH 7 ± 0.1) containing 5 mM $[Fe\ (CN)_6]^{3-/4-}$. In the CV curve of PHA, no significant redox peak was observed indicating PHA none electroactive. In the CV curves of PHA/AuNPs/ITO, the peaks shifted towards more positive potential. The AuNPs modified electrode having improved electrochemical behavior can be attributed to the increased charge propagation within the electrode due to the incorporation of AuNPs. In the presence of artemisinin at 0.01 μg mL^{-1}, the effect of pH on the peak current at the range of 5.5 to 8.5 was investigated. The electrodes showed a maximum peak current and good redox potential at pH 7 ± 0.1 of PBS. Therefore, to maximize current signal of biosensors, a PBS buffer solution (pH 7 ± 0.1) was optimized as the optimum solution. The redox current change represented increasing scan rate effecting a steady increase in peak current and peak potential. Also, uniformity of current function, IP/AC1/2 was observed from the plot.

ip (μA) = 8.3 × 1/2 (mVs^{-1}) + 9.1; $r^2 = 0.992$; n = 8 (PHA/AuNPs/ITO)
ip (μA) = 0.12 × 1/2 (mVs^{-1}) + 9.8; $r^2 = 0.996$; n = 8 (PHA/AuNPs/HRP/ITO)

There was no significant peak for the PHA therefore no slope was assigned, the addition of AuNPs into PHA could produce sharp anodic and cathodic peaks slope 0.122 μA $(mV^{-1})^{1/2}$ and 8.31 μA $(mV^{-1})^{1/2}$, respectively. This is because of the highly conducting AuNPs facilitates the oxido-reduction inside the electrodes. After HRP immobilization, slopes were decreased to 0.106 μA $(mV^{-1})^{1/2}$ and 0.168 μA $(mV^{-1})^{1/2}$, respectively due to the effective immobilization of retaining the bioactivity of PHA/AuNPs/ITO. CV's relative variation at constant scan rate (5 mVs^{-1}) in PBS (50 mM, pH 7.01) containing 5 mM $[Fe\ (CN)_6]^{3-/4-}$ for three electrodes (a) PHA/ITO (b) PHA/AuNPs/ITO and (c) PHA/AuNPs/HRP/ITO proved the functionality. In the PHA/ITO electrode, no significant redox peak was observed as PHA acts as a barrier in electron transport. The CV curve of PHA/AuNPs/ITO revealed significant redox peaks with a near rectangular shape. Electrode's improved electrochemical behavior was due to the addition of AuNPs facilitating charge transfer within the system. In PHA/AuNPs/HRP/ITO the peak current decreased in comparison to PHA/AuNPs/ITO which might be due to slow redox process following immobilization of HRP.

14.17.2 EIS STUDY OF BIOSENSOR AUNPS

EIS was carried out on (a) PHA/ITO, (b) PHA/AuNPs/ITO and (c) PHA/AuNPs/HRP/ITO electrode for the determination of relative change in surface resistance at '0' potential. The RCT surface charge transfer resistance Rp, the polarization resistance obtained at zero potential and Cd are double layer capacitance at the electrode/solution interface. The frequency associated with maximum Z (imaginary impedance) and RCT were used to calculate Cd using the following equation:

$$\mathrm{RCTCd} = 1/2f_{max}$$

Here, f_{max} is the frequency at which maximum Z obtained.

From the Nyquist plot it was seen that the Nyquist diameter of the electrode deposited with PHA is much larger than that of the PHA/AuNPs electrode, which manifested that the formation of PHA layer on the electrode surface, and it in some degree blocked the redox probe of $[Fe(CN)_6]^{3-/4-}$ contacting with the electrode surface. This clearly indicated that the AuNPs were not only self-assembled on the outer surface of the PHA, but also partly distributed within the lm as tiny conduction centers which facilitated the electron-transfer and evidenced more conductive. However, AuNPs act as electron-conducting tunnel which can greatly promote the direct electron transfer. This change in the capacitance suggested that AuNPs were successfully incorporated into the PHA films and this is further supported by CV study which shows a drastic enhancement of current in PHA/AuNPs as compared to PHA. It has also been noticed in the impedance spectrum of the resulting electrode immobilized with HRP that the increase in RCT was due to the dielectric and insulating features at the electrodes electrolyte interface and was associated with successive immobilization of HRP. The change in RCT after HRP immobilization was also might be due to the hindrance of the macromolecular structure of

HRP to the electron transfer, and it also confirmed the successful immobilization of HRP.

14.17.3 SEM STUDY OF BIOSENSOR WITH AUNPS

To get insight into the surface morphology of the working electrodes SEM micrograph was investigated. The non-porous and smooth surface of PHA agrees with CV studies confirming non-conductivity of PHA being non-electroactive in nature. The electrode surface consists of an uneven porous matrix of PHA, embedded with AuNPs of around 66.67 nm increasing the conductivity of the film. PHA/AuNPs could provide a matrix for the successful immobilization of the HRP enzyme. Lin et al. (2007) developed a disposable biosensor for rapid detection of H_2O_2 by incorporating HRP enzyme in Auchitosan membrane to modify ITO electrode. The presence of Auchitosan membrane plays an important role in the immobilization of HRP enzyme and facilitates immobilization to expose activity sites of the enzyme making the substrate more easily accessible to the enzyme, causing a good ampere-metric response of the biosensor. AuNPs have attributed increase in roughness of the surface area and caused effective sensitivity of the biosensor.

14.17.4 DETECTION OF ANTIMALARIA DRUG-ARTEMISININ IN PHA/AUNPS/HRP/ITO SAMPLE

The applicability of PHA/AuNPs/HRP/ITO bio-electrode-based electrochemical biosensor was studied in bulk and spiked human serum. The current is proportional to the concentration of artemisinin over a convenient range. The precision was estimated for 0.01–0.08 μgmL^{-1} of the drug using the calibration curve and the standard addition method. The calibration equation parameters and related validation data have been applied. Recovery was performed using the standard addition method and recovery values of artemisinin (0.01 $\mu g\ mL^{-1}$) in spiked serum was 95.41% with % RSD of 1.05, n = 5. It demonstrated repeatability and reproducibility along with precision and confirms to be a potential alternative for the direct determination of artemisinin in pharmaceutical formulations and spiked biological fluids. The strategy described here has many attractive advantages: simplicity, rapidity, and low cost could be a useful method for pharmaceutical analysis. The electrochemical behavior of artemisinin is an irreversible process controlled by diffusion. The method proposed has a distinct advantage over other existing methods regarding sensitivity, selectivity, time saving and detectability. Furthermore, in HPLC and GC methods for the determination of artemisinin, sensitivity, and low detection were found to be 5, 50, and 30 $ngmL^{-1}$, by the proposed method the same could be estimated up to a level of 0.0035 $\mu g\ mL^{-1}$, on application to spiked human serum detection up to 0.0036 μgmL^{-1} proved it to be more sensitive to HPLC (Jastrebova et al., 1998; Peng et al., 2006; Amponsaa-Karikari et al., 2006). The proposed method could be a good analytical alternative in comparison to chromatographic methods for the detection of artemisinin.

KEYWORDS

- **artemisinin**
- **Fourier-transform-infrared**
- **gold nanoparticles**
- **hyperbranched epoxy**
- **modified hyperbranched epoxy**
- **nanoparticles**

REFERENCES

Amponsaa-Karikari, A., et al., (2006). Determination of artemisinin in human serum by high-performance liquid chromatography with on-line UV irradiation and peroxyoxalate chemiluminescence detection. *Biomed Chromatogr., 20*, 1157–1162.

Bharali, P., & Konwar, B. K., (2011). Production and physicochemical characterization of a biosurfactant produced *by Pseudomonas aeruginosa* OBP1 isolated from petroleum sludge. *Appl. Biochem. Biotechnol., 164*(8), 1444–1460.

Green, M. D., et al., (1995). Chemiluminescent detection of artemisinin novel endoperoxide analysis using luminol without hydrogen peroxide. *J. Chromatogr A, 695*, 237–242.

Jastrebova, J., et al., (1998). On-line deoxygenation for reductive electrochemical detection of artemisinin and dihydroartemisinin in liquid chromatography. *Analyst, 123*, 313–317.

Kausik, A., et al., (2008). Fabrication and characterization of polyaniline-ZnO hybrid nanocomposite thin films. *J. Nanosci. Nanotechnol., 8*, 1757–1761.

Kim, D. Y., & Rhee, Y. H., (2003). Biodegradation of microbial and synthetic polyesters by fungi. *Appl. Microbiol. Biotechnol., 61*, 300–308.

Lin, J., et al., (2007). Disposable biosensor based on enzyme immobilized on Au–chitosan-modified indium tin oxide electrode with flow injection amperometric analysis. *Anal Biochem., 360*, 288–293.

Peng, C. A., et al., (2006). Direct analysis of artemisinin from *Artemisia annua* L. using high-performance liquid chromatography with evaporative light scattering detector, and gas chromatography with flame ionization detector. *J. Chromatogr. A, 1133*, 254–258.

Philip, S., et al., (2007). Polyhydroxyalkanoates: Biodegradable polymers with a range of applications. *J. Chem. Technol. Biotechnol., 82*, 233–247.

Roy, B., Bharali, P., Konwar, B. K., & Karak, N., (2013). Modified hyperbranched epoxy/clay nanocomposites: Anti-fungal, thermal and biodegradation study. *Colloids and Surfaces B: Biointer., 102*, 450–456.

Roy, B., Bharali, P., Konwar, B. K., & Karak, N., (2013). Silver-embedded modified hyperbranched epoxy/clay nanocomposites as antibacterial materials. *Bioresource Tech., 127C*, 175–180.

Saha, S., et al., (2010). Production of silver nanoparticles by a phytopathogenic fungus *Bipolaris nodulosa* and its antimicrobial activity. *Digest J. of Nanomat. Biostruct., 5*, 887–895.

Solomon, S. D., et al., (2007). Synthesis and study of silver nanoparticles. *J. Chem. Educ., 84*, 322–325.

Udapudi, B. B., et al., (2012). Synthesis and characterization of silver nanoparticles. *International Journal of Pharmacy and Biological Sciences, 2*, 10–14.

Zhang, L., et al., (2005). Attachment of gold nanoparticles to glassy carbon electrode and its application for the direct electrochemistry and electrocatalytic behavior of hemoglobin. *Biosens. Bioelectron., 21*, 337–345.

Conclusion

1. Three different PHA-producing bacterial isolates, BPC1, BPC2, and BP2, were recovered from the crude oil-contaminated soil samples collected from the Siwasagar and Jorhat districts of Assam.
2. The waste glycerol byproduct of the kitchen chimney dump lard (KCDL) was found to be a good carbon source for the highest PHA accumulation in all three bacterial strains in comparison to other carbon sources used.
3. Based on FTIR, GCMS, and ^{1}H and ^{13}C NMR, their characterization led to the identification of the biopolymers isolated from *P aeruginosa* JQ796859, *B circulans* MTCC8167, and *P aeruginosa* JQ866912 to be poly (3-hydroxyvalerate) co-(5-hydroxydecenoate) (P-3HV-5-HDE), poly-3-hydro xybutyrate-co-3-hydroxyvalerate (P-3HB-3HV) and poly-3-hydroxyvale-rate-co-5-hydroxydecenoate-co-3-hydroxyoctadecenoate(P-3HV-5HDE-3HODE), respectively.
4. By incorporating the metal oxide nanoparticles with biopolymer, the intensity of the emission peak could be increased. The resulting nanocomposites could be used for further application as biosensors.
5. A selective and sensitive PHA/AuNPs/HRP/ITO biosensor-based nanocomposite probe is developed for the direct determination of artemisinin in bulk and spiked human serum. The proposed method has a distinct advantage over other existing methods regarding sensitivity, selectivity, time-saving, and minimum detectability.
6. The resultant specific 540-bp PCR product has represented the presence of mcl biosynthesis genes phaC1/C2 in the bacterial strains JQ796859 and JQ866912 of *P aeruginosa.*

Further Reading

Agus, J., et al., (2006). Molecular weight characterization of poly [(R)-3-hydroxybutyrate] synthesized by genetically engineered strains of *Escherichia coli*, *Polym. Degrad. Stab., 91*, 1138–1146.

Aldor, I. S., & Keasling, J. D., (2003). Process design for microbial plastic factories: Metabolic engineering of polyhydroxyalkanoates. *Curr. Opin. Biotechnol., 14*, 475–483.

Choi, G. G., et al., (2004). Enzymatic and non-enzymatic degradation of poly (3-hydroxybutyrate-co-3-Hydroxyvalerate) copolyesters produced by *Alcaligenes* spp MT-16. *J, Microbiol., 42*, 346–352.

Cornibert, J., & Marchessault, R. H., (1972). Physical properties of poly-hydroxybutyrate. IV. Conformational analysis and crystalline structure. *J. Mol. Biol., 71*, 735–756.

Gao, Y., et al., (2006). Improvement of mechanical properties of poly (DL-lactide) films by blending of poly (3-hydroxybutyrate-co-3-hydroxyhexanoate). *Eur. Polym. J., 42*, 764–775.

Griebel, R., et al., (1968). Metabolism of poly (β-hydroxybutyrate). I. Purification, composition, and properties of native poly (β-hydroxybutyrate) granules from *Bacillus megaterium*. *Biochemistry, 7*, 3676–3681.

Kim, B. S., et al., (1994). Production of poly (3-hydroxybutyric acid) by fed-batch culture of *Alcaligenes eutrophus* with glucose concentration control. *Biotechnol. Bioeng., 43*, 892–898.

Kunze, C., et al., (2006). *In vitro* and in vivo studies on blends of isotactic and atactic poly (3-hydroxybutyrate) for development of a dura substitute material, *Biomaterials, 27*, 192–201.

Lee, K. M., & Gilmore, D. F., (2006). Modeling and optimization of biopolymer (polyhydroxyalkanoates) production from ice cream residue by novel statistical experimental design. *Appl. Biochem. Biotechnol., 133*, 113–148.

Lillo, J. G., and Rodriguez-Valera, F., (1990). Effects of culture conditions on poly (β-hydroxybutyric acid) production by *Haloferax mediterranei*. *Appl. Environ. Microbiol., 56*, 2517–2521.

Lu, Y. P., (2007). Advance on the production of polyhydroxyalkanoates by mixed cultures. *Front. Biol. China, 2*, 21–25.

Majone, M., et al., (1996). Influence of storage on kinetic selection to control aerobic filamentous bulking. *Water Sci. Technol., 34*, 223–232.

Nakashita, H., et al., (1999). Production of biodegradable polyester by a transgenic tobacco. *Biosci. Biotechnol. Biochem., 63*, 870–874.

Omar, S., et al., (2001). Optimization of cell growth and poly (3-hydroxybutyrate) accumulation on date syrup by a Bacillus megaterium strain. *Biotechnol. Lett., 23*, 1119–1123.

Padermshoke, A., et al. (2005). Melting behavior of poly (3- hydroxybutyrate) investigated by two-dimensional infrared correlation spectroscopy. *Spectrochim. Acta, Part A–Mol. Biomol. Spectrosc., 61*, 541–550.

Phukan, P., (2013). *Biochemical and Molecular Genetic Assessment of Bacterial Biopolymer* (PhD Thesis), Tezpur University (Central), Napaam, Tezpur, Dist Sonitpur, Assam, India.

Quillaguaman, J., et al., (2005). Poly (b-hydroxybutyrate) production by a moderate halophile, *Halomonas boliviensis* LC1 using starch hydrolysate as substrate. *J. Appl. Microbiol., 99*, 151–157.

Quillaguaman, J., et al., (2010). Synthesis and production of polyhydroxyalkanoates by halophiles: Current potential and future prospects. *Appl. Microbiol. Biotechnol., 85*, 1687–1696.

Randriamahefa, S., et al., (2003). Fourier transform infrared spectroscopy for screening and quantifying production of PHAs by *Pseudomonas* grown on sodium octanoate. *Biomacromolecules, 4*, 1092–1097.

Steinbuchel, A., & Fuchtenbusch B (1998). Bacterial and other biological systems for polyester production. *Trends Biotechnol., 16*, 419–427.

Suzuki, T., et al., (1986a). Kinetics and effects of nitrogen source feeding on production of poly-β-hydroxy butyric acid by fed-batch culture. *Appl. Microbial. Technol., 24*, 366–369.

Suzuki, T., et al., (1986b). Mass production of poly-β-hydroxybutyric acid by fed-batch culture with control carbon/nitrogen feeding. *Appl. Microbial. Biotechnol., 24*, 370–374.

Tsz-Chun, M., et al., (2005). Microbial synthesis and characterization of physiochemical properties of polyhydroxyalkanoates (PHAs) produced by bacteria isolated from activated sludge obtained from the municipal wastewater works in Hong Kong. *Appl. Biochem. Biotechnol., 122*, 731–740.

Wang, F., & Lee, S. Y., (1997). production of poly (3-hydroxybutyrate) by fed-batch culture of filamentation-suppressed recombinant *Escherichia coli*. *Appl. Environ. Microbiol., 63*, 4765–4769.

Wieczorek, R., et al., (1995). Analysis of a 24-kilodalton protein associated with the polyhydroxy alkanoic acid granules in *Alcaligenes eutrophus*. *J. Bacteriol., 177*, 2425–2435.

Yong, K. H., (2008). *Biosynthesis and Characterization of Polyhydroxyalkanoates by Locally Isolated Chromobacterium sp.* USM2, PhD Thesis.

Index

C

D

E

F

G

T

For Product Safety Concerns and Information please contact our EU representative GPSR@taylorandfrancis.com Taylor & Francis Verlag GmbH, Kaufingerstraße 24, 80331 München, Germany